Ecological Consequences of the Second Indochina War

SIPRI

Stockholm International Peace Research Institute

SIPRI is an independent institute for research into problems of peace and conflict, especially those of disarmament and arms regulation. It was established in 1966 to commemorate Sweden's 150 years of unbroken peace.

The institute is financed by the Swedish Parliament. The staff, the Governing Board and the Scientific Council are international. As a consultative body, the Scientific Council is not responsible for the views expressed in the publications of the Institute.

SIPRI
Stockholm International Peace Research Institute
Sveavägen 166, S-113 46 Stockholm, Sweden
Cable: Peaceresearch, Stockholm
Telephone: 08-15 09 40

Ecological Consequences of the Second Indochina War

SIPRI

Stockholm International Peace Research Institute

Almqvist & Wiksell
International
Stockholm, Sweden

First published by the

Stockholm International Peace Research Institute

in cooperation with

Almqvist & Wiksell International
26 Gamla Brogatan, S-111 20 Stockholm, Sweden

ISBN 91-22000-62-3

Printed in Sweden by
Tryckindustri AB, Solna 1976

PREFACE

Environmental considerations are today intruding themselves upon a wide range of human endeavours with ever increasing urgency. Not least among these endeavours are the military ones.

Two important international conferences have already taken up the question of the environmental impact of modern warfare. At the second session of the Diplomatic Conference on Humanitarian Law, held in Geneva from February to May 1975, proposals were put forward which, if adopted by the States, would introduce as a new basic rule of warfare a prohibition of "methods or means of warfare which are intended or may be expected to cause widespread, long-term, and severe damage to the natural environment". In August 1975 the United States and the Soviet Union presented identical proposals at the Conference of the Committee on Disarmament in Geneva to prohibit various means of environmental modification as a means of warfare. In addition to the prohibition of a number of exotic means (some of them of dubious feasibility) the 1974 Soviet proposals would also outlaw means of warfare which cause a disturbance of the "dynamics, composition or structure of the earth, including its biota, lithosphere, hydrosphere, and atmosphere . . . so as to cause such effects as . . . an upset in the ecological balance of a region, or changes in weather patterns".

The recently concluded Second Indochina War provides an important case history of the environmental impact of modern limited conventional warfare. It provides the main source of information and theme for the present monograph.

This monograph was written by Arthur H. Westing, Professor of Botany at Windham College in Putney, Vermont, USA, while a visiting researcher at SIPRI. It is based on a longer work in progress on *Warfare and the Environment,* which, unlike the present monograph, will also pay attention to the social, economic and public health aspects of environmental destruction. The author made four investigative trips to the war zones of Indochina between 1969 and 1973.

February 1976

Frank Barnaby
Director

CONTENTS

TABLES AND MAPS

Chapter 1. The Second Indochina War

TABLES

MAPS

Chapter 2. High-explosive munitions

TABLES

Chapter 3. Anti-plant chemicals

TABLES

Chapter 5. Miscellaneous weapons and techniques

TABLES

CONVENTIONS, ABBREVIATIONS AND UNITS OF MEASURE

Conventions

Insofar as possible, scientific names of plants conform to G. Lawrence (1951) and of mammals to E. Walker *et al.* (1964). Chemical nomenclature follows that of Stecher *et al.* (1968) or, secondarily, of Weast (1974:B-C).

The following conventions are used in the tables:

..	Datum not available
–	Datum nil or negligible
( )	Datum uncertain

Abbreviations

The following abbreviations are used in the text. Omitted, however, are the international symbols for the chemical elements (such as As = arsenic), and the international units of measure, for which see below.

B-52	Boeing Co. "Stratofortress" bomber aircraft, Model B-52
2,4-D	2,4-dichlorophenoxyacetic acid
M.R.	Military Region
2,4,5-T	2,4,5-trichlorophenoxyacetic acid
°	Degree
/	Per

Units of measure

Units of measure follow the *Système International d'Unités* (SI) (Page & Vigoureux, 1972). Conversions are from Weast (1974:F:282-304).

a	= are = 10^2 square metres = 1 076.39 square feet
c-	= centi- = 10^{-2} x
cm	= centimetre = 10^{-2} metre = 0.393 701 inch
d	= day = 86 400 seconds
°C	= degree Celsius (To obtain temperature in degrees Fahrenheit, multiply by 1.8 and then add 32)
g	= gram = 10^{-3} kilogram = 2.204 62 x 10^{-3} pound
g/m³	= gram per cubic metre = 8.345 40 x 10^{-6} pound per US gallon = 10.022 4 x 10^{-6} pound per British gallon
h	= hour = 3 600 seconds
h-	= hecto- = 10^2 x
ha	= hectare = 10^4 square metres = 10^{-2} square kilometre = 2.471 05 acres
J	= joule = 0.238 846 calorie
k-	= kilo- = 10^3 x
kg	= kilogram = 2.204 62 pounds
kg/ha	= kilogram per hectare = 0.892 179 pound per acre
kg/m	= kilogram per metre = 0.671 969 pound per foot

kg/m^3 = kilogram per cubic metre = 8.345 40 x 10^{-3} pound per US gallon = 10.022 4 x 10^{-3} pound per British gallon = 1.685 55 pounds per cubic yard
km = kilometre = 10^3 metres = 0.621 371 mile
km^2 = square kilometre = 10^2 hectares = 247.105 acres =0.386 102 square mile
kPa = kilopascal = 9.869 23 x 10^{-3} atmosphere = 0.145 038 pound per square inch
l = litre = 10^{-3} cubic metre = 0.264 172 US gallon = 0.219 969 British gallon
l/ha = litre per hectare = 0.106 907 US gallon per acre = 0.089 0185 British gallon per acre
m = metre = 3.280 84 feet
m- = milli- = 10^{-3} x
m^2 = square metre = 10.763 9 square feet
m^2/kg = square metre per kilogram = 4.882 43 square feet per pound
m^3 = cubic metre = 10^3 litres = 264.172 US gallons = 219.969 British gallons = 1.307 95 cubic yards = 138 board feet
m^3/ha = cubic metre per hectare = 55.846 6 board feet per acre
m^3/kg = cubic metre per kilogram = 0.593 276 cubic yard per pound
mg = milligram = 10^{-6} kilogram = 2.204 62 x 10^{-6} pound
mg/ha = milligram per hectare = 892.179 x 10^{-9} pound per acre
mg/kg = milligram per kilogram = part per 10^6 parts (ppm), by weight
mg/m^3 = milligram per cubic metre = 8.345 40 x 10^{-9} pound per US gallon = 10.022 4 x 10^{-9} pound per British gallon
min = minute = 60 seconds
ml = millilitre =10^{-6} cubic metre = 61.023 7 x 10^{-3} cubic inch
mm = millimetre = 10^{-3} metre = 0.039 3701 inch
ms = millisecond = 10^{-3} second
m/s = metre per second = 3.6 kilometres per hour = 2.236 94 miles per hour
M- = Mega- = 10^6 x
MJ = Megajoule = 10^6 joules = 238 846 calories
MJ/kg = Megajoule per kilogram = 108 339 calories per pound
μ- = micro- = 10^{-6} x
μm = micrometre = 10^{-6} metre = 39.370 1 x 10^{-6} inch
n- = nano- = 10^{-9} x
ng = nanogram = 10^{-12} kilogram = 2.204 62 x 10^{-12} pound
ng/kg = nanogram per kilogram = part per 10^{12} parts, by weight
Pa = pascal = 9.869 23 x 10^{-6} atmosphere = 145.038 x 10^{-6} pound per square inch
s = second
t = tonne = 10^3 kilograms = 1.102 31 US (short) tons = 0.984 207 British (long) ton

Chapter 1. The Second Indochina War

Where indicated thus,[1] the reader is referred to the notes on page 10.

I. *Introduction*

Rightly or wrongly, the Second Indochina War[1] has introduced into the military lexicon the new word "ecocide".[2] The Second Indochina War was by no means the first in which ecological or environmental disruption has occurred, or even the first one in which such damage was the result of conscious policy. However, this war does stand out in modern history as one in which intentional anti-environmental actions were a major component of the strategy and tactics of one of the adversaries, one in which such actions were systematically carried out for many years and over large areas.

This military assault on the natural resources of Indochina comes at a time in human history when man has begun to realize the awesome dimensions of his routine demands on and abuses of nature. But most of all, it comes at a time when the world is awakening to a recognition of the inextricable interdependence of man and nature, to man's obligate dependence for his well-being and very survival upon a finite and increasingly misused globe.

As a result, it becomes crucial to examine the Second Indochina War as a case study of modern environmental abuse, the aim of the present monograph. Following brief descriptions of both Indochina (section II) and the war (section III) as these apply to the theme of the study, the various means of environmental assault are next examined. These include high-explosive munitions (chapter 2), anti-plant chemicals (chapter 3), mechanized landclearing (chapter 4) and a number of miscellaneous weapons and techniques (chapter 5). Those studies of individual anti-environmental weapons and techniques are, in turn, followed by a discussion of the overall implications of such assault, first for nature (chapter 6) and then for man (chapter 7).

II. *The theatre of war*

Indochina – South Viet-Nam, North Viet-Nam, Cambodia, and Laos – occupies a major fraction of the Southeast Asian Peninsula.[3] It is an area of forbidding mountains, gentle hills, and flat plains situated wholly between the equator and the Tropic of Cancer (map 1.1). It falls between about 8 1/2° and 23 1/2° north latitude and between about 100° and 109 1/2° east longitude. Some of the mountain peaks rise above 2 000 m and a few over 3 000 m.

Although Indochina's 76 x 10^6 ha are rather evenly divided among the four countries, its population is much less evenly apportioned (table 1.1). Nation-

Map 1.1. Indochina: an outline map of the countries

wide population densities range from a high of 119/km² in North Viet-Nam to a low of 12/km² in Laos. The annual rate of population increase falls between 2 and 3 per cent. Roughly nine out of 10 of the inhabitants of Indochina are (or were before the war) farmers or fishermen. The vast majority are concentrated in the relatively flat areas. Approximately 5 to 10 per cent of the re-

Table 1.1. Indochina: areas and populations

Region	Area[a] 10^3 ha	Population[b] 10^3	Density No./km^2
South Viet-Nam	17 326	17 633	102
Military Region I	2 812	3 075	109
Military Region II	7 696	3 086	40
Military Region III	3 021	4 858	161
Without Saigon		2 358	78
Military Region IV	3 797	6 614	174
North Viet-Nam	16 406	19 446	119
Cambodia	18 104	6 649	37
Laos	23 680	2 891	12
Total	**75 516**	**46 619**	**62**

Notes and sources:

[a] Areas: The area of South Viet-Nam is taken from Engineer Agency . . . (1968:87–89) and is supported by Tung (1967:1). The areas of the Military Regions are also taken from Engineer Agency . . . (1968:87–89), being additions of the appropriate provinces. The area of North Viet-Nam is taken from H. Smith *et al.* (1967a:11). The area of Cambodia is from Whitaker *et al.* (1973:7). The area of Laos is from Whitaker *et al.* (1972:11) and is supported by Engineer Agency . . . (1968: 87–89).

[b] Populations: The populations must all be considered as approximations inasmuch as none of the countries involved has ever undergone a regular census. The datum for Laos is particularly unreliable. All of the figures presented here have been adjusted to the year 1969, the midpoint of the Second Indochina War. The population of South Viet-Nam is calculated from the 1965 value of 16 100 x 10^3 given by H. Smith *et al.* (1967b:59) using an annual increase of 2.30 per cent, the average of those given by H. Smith *et al.* (1967b:59) and Engineer Agency . . . (1968:87–89). This is partitioned among the Military Regions in accordance with the proportions derived from the province data given by Engineer Agency . . . (1968:87–89). The population of North Viet-Nam is calculated from the 1960 values of 15 917 x 10^3 given by H. Smith *et al.* (1967a:26–27) using the annual increase of 2.25 per cent provided in the same source. The population of Cambodia is calculated from the 1962 value of 5 729 x 10^3 given by Engineer Agency . . . (1968:87–89) using the annual increase of 2.15 per cent provided in the same source. The population of Laos is calculated from the 1970 value of 2 962 x 10^3 provided by Whitaker *et al.* (1973:20) using the annual increase of 2.45 per cent provided in the same source as well as by Engineer Agency . . . (1968:87–89).

gional population is composed of a diverse number of primitive ethnic minority groups. Many of these so-called Montagnards lead a semi-nomadic tribal existence in the highlands of Indochina.

The climate of Indochina is generally hot, rainy and humid. The southeast summer monsoons bring high temperatures and a deluge of rain. The northeast winter monsoons are somewhat less rainy but only slightly cooler. In Saigon (elevation 9 m) the average annual rainfall is 1 944 mm, average annual temperature 27°C, and average annual relative humidity 80 per cent. Somewhat further north and higher up, in Pleiku (elevation 800 m), the comparable figures are, respectively, 2 451 mm, 22°C and 84 per cent (Engineer Agency . . . , 1968:8–9).

Indochina is dominated by two powerful rivers, the Red river to the north and the Mekong river to the south. Their immense deltas are covered by rice (*Oryza sativa*, Gramineae) paddies that from the air appear to be vast patchwork quilts. The rugged highlands, covering some two-thirds of the region, are characterized by their own patchwork of countless tiny (several hectare) plots. These have been carved out of the jungle for centuries, perhaps millennia, by the primitive Montagnards who roam these largely uncharted mountains practising shifting slash-and-burn agriculture. Some of the patches support crops. Some are too impoverished to support anything but low weeds. However, most are covered by secondary forest growth in various stages of successional development. The soils of Indochina range from rich alluvial soils in the delta regions to impoverished red latosols in the highland regions. The upland forest soils are usually acid in reaction, low in humus, heavily leached and with very low nutrient-holding capacity. The propensity to at least rapid laterization or induration upon complete exposure appears to be a rather rare and scattered phenomenon.

Although there is no readily available breakdown into land-use categories for Indochina as a whole, some estimates for South Viet-Nam can be presented (table 1.2). Of South Viet-Nam's 17 x 10^6 ha of land, approximately 57 per cent is covered by a diversity of upland (inland) forests, 1 per cent by rubber (*Hevea brasiliensis*, Euphorbiaceae) plantations, 2 per cent by coastal mangrove forests (swamps), 14 per cent by paddy (wet) rice (*Oryza sativa*, Gramineae), 3 per cent by dry-field crops, and the remaining 23 per cent by a miscellany of types (including grasslands or savannas, the so-called Plain of Reeds vegetational type, open water and urban areas). Although the immense array of South Viet-Namese higher plant and animal species appears to have been quite well catalogued, there exist as yet no adequate ecological analyses and no systematic timber volume inventories.

The several tree-covered areas of South Viet-Nam mentioned above add up to a little more than 10 x 10^6 ha. Of this combined area, about 56 per cent can be categorized as dense (closed) upland (inland) forest (the layman's "jungle"), much of it in various stages of succession. The dense upland forest type contains a bewildering diversity of dicotyledonous trees, lianas, epiphytes and herbs as well as some monocotyledons, ferns and so forth. The tree species vary in height, usually forming two and occasionally three rather indistinct strata (storeys). The upper canopy usually attains a height of 20 m to 40 m. The dominant plant family is the Dipterocarpaceae which is represented by at least 30 major species in the genera *Dipterocarpus*, *Anisoptera*, *Hopea* and *Shorea*. Another important genus is *Lagerstroemia* in the family Lythraceae. There are also a number of important genera of Leguminosae (for example, *Erythrophleum*), Guttiferae and Meliaceae. The forests support a rich fauna, particularly in their upper storeys.

Table 1.2. South Viet-Nam: vegetational types

Vegetational type	Area 10^3 *ha*
Dense forest	5 800
Pure dense forest	4 500
Pure plus secondary dense forest	600
Secondary dense forest	700
Open (clear) forest	2 000
Bamboo brake	800
Mangrove forest (swamp)	500
True mangrove	300
Rear mangrove	200
Rubber plantation	100
Pine forest	100
Miscellaneous woody (brush, etc.)	1 100
Woody subtotal	**10 400**
Paddy (wet) rice	2 500
Field crops (upland rice, etc.)	500
Agricultural subtotal	**3 000**
Miscellaneous	3 926
Total	**17 326**

Notes and sources:

(*a*) The total area of South Viet-Nam is from table 1.1. Areas of the vegetational types are derived from a South Viet-Namese type map (National Geographic Service, 1969) by the method of tracing the types and weighing the cutouts. They are verified where possible by the data of Tung (1967). The "Miscellaneous" category includes the so-called Plain of Reeds vegetational type (500 x 10^3 ha), grassland or savanna (1 000 x 10^3 ha), and miscellaneous herbaceous types. It also includes open water (200 x 10^3 ha) and various non-vegetational land categories.

(*b*) To provide at least an indication of how the forest lands of South Viet-Nam are partitioned among the Military Regions, it is possible to cite the comparable data compiled by the South Viet-Namese forest service for the 5 900 x 10^3 ha of commercially exploitable forests of the country: M.R. I 26 per cent; M.R. II 52 per cent; M.R. III 18 per cent; and M.R. IV 4 per cent (Tân, 1971).

(*c*) Rubber = *Hevea brasiliensis*, Euphorbiaceae. Pine = *Pinus*, Pinaceae. Rice = *Oryza sativa*, Gramineae.

Thus, as has been suggested above, the upland forests of Indochina are a confusing conglomeration of what appears to be primary forest interspersed with secondary forest in all stages. Moreover, the forests have been subjected over the years and centuries to varying intensities of exploitation for timber, firewood and miscellaneous products. And of course, many years of war have left their mark as well, in a variety of obvious and subtle ways.

The estuaries along the coastal fringe particularly of southern Indochina support a lowland forest type known as mangrove. The mangrove habitat is thus found primarily in South Viet-Nam, where it occupies a combined area estimated to be in the neighbourhood of 0.5×10^6 ha. In contrast with the dense upland forest type with its several thousand plant species, the mangrove type is composed of a mere several dozen. This handful of species has in common the ability to become established and survive in a mucky soil which is periodically inundated with salt water. The description which follows is justified by the especially high level of war damage suffered by this particular type (see page 38).

The dominant mangrove vegetation consists of several species of small trees, mostly 3–15 m high, primarily in the genera *Rhizophora* (Rhizophoraceae), *Avicennia* (Verbenaceae) and *Bruguiera* (Rhizophoraceae) – all so-called mangroves. As soil deposition extends the coastline slowly out into the sea, the most common pioneer of this virgin land is *Sonneratia* (Lythraceae), which in turn prepares the site for a subsequent invasion by *Avicennia*. *Rhizophora* is likely to be next in this succession, to be followed by *Bruguiera* which extends back to the limit of daily tidal wash.

Through time, the soil level builds up beyond the reach of flood tide, and the more or less concentric zones just described – which together comprise the so-called true mangrove type – give way to a new community known as rear (or back) mangrove. The rear mangrove type is dominated by a tree species of *Melaleuca* (Myrtaceae). Of the total South Viet-Namese mangrove forests, perhaps 60 per cent consists of true mangrove and the remaining 40 per cent of rear mangrove (table 1.2). The U Minh forest in Kien Giang and An Xuyen provinces is South Viet-Nam's most extensive example of rear mangrove (see map 1.2).

The vegetative component of the mangrove community is of minor regional importance for primary forest products such as timber, but provides substantial amounts of firewood, charcoal, tannin, thatch (from *Nipa fruticans,* Palmae), and other secondary forest products. This tropical estuarine channel-dissected habitat is also of major regional importance as the nursery or breeding grounds for numerous saltwater and freshwater fish and crustaceans.

III. *The war strategy and tactics*

The Second Indochina War – the longest and perhaps most costly of any US war – is a difficult one to summarize satisfactorily.[4] Overt US involvement began quite subtly around 1961, built up slowly to a rather diffuse climax around 1968, and then trailed off inauspiciously around 1973 (table 1.3). During this extended period US armed forces attempted in South Viet-Nam to cope with a persistent and mobile enemy guerrilla force numbering perhaps 600 000 (S. Adams, 1975). Throughout the war the United States maintained physical, on-the-ground control of only a tiny fraction of South Viet-Nam. That portion,

Map 1.2. South Viet-Nam: an outline map of the Military Regions and War Zones

LiberKartor Stockholm 1976

however, contained in its fragments the various important urban areas of the country and a large majority of its population.

Elsewhere in Indochina the situation was rather different. The USA attempted no on-the-ground control of North Viet-Nam, and in Cambodia and Laos essentially only their capital cities. The US wars against North Viet-

Table 1.3. Some measures of US effort during the Second Indochina War: a breakdown by year

Year	Forces committed[a] 10^3	Killed in action[b]	Proportion killed in action *No per 10^3*	Munitions expended[c] *10^6 kg*	Herbicides expended[d] *m^3*
1961	..	..	..	..	..
1962	9	42	5	..	65
1963	15	78	5	..	283
1964	16	147	9	..	1 066
1965	60	1 369	23	286	2 516
1966	268	5 008	19	998	9 599
1967	449	9 377	21	1 965	19 394
1968	535	14 589	27	2 696	19 264
1969	539	9 414	17	2 561	17 257
1970	415	4 221	10	1 970	2 873
1971	239	1 381	6	1 453	38
1972	47	300	6	1 792	–
1973	..	219	..	543	–
Total	**2 592**	**46 145**	**18**	**14 265**	**72 354**

Notes and sources:

[a] The number of US military forces committed is from a US Department of Defense release of 19 March 1973 and represents numbers in South Viet-Nam on 30 June of each year.

[b] The number killed in action is from a US Department of Defense release of 28 November 1973 and represents all deaths resulting from actions by hostile forces. These data, therefore, do not include the 1 014 missing in action nor do they include the 10 320 military deaths not the result of actions by hostile forces.

[c] Munition expenditures are from table 2.1.

[d] Herbicide expenditures are from table 3.2.

Nam, Cambodia and Laos – mounted mostly from the air – were in large measure ancillary to and in support of its war in South Viet-Nam. Indeed, as shattering as the war was to each of the countries of Indochina, it was South Viet-Nam that absorbed the major fury of the US assault, particularly its Military Region III (see table 1.4 and map 1.2).

The United States was loath to commit its army to the sustained ground war (with its attendant high casualties) necessary to achieve a military victory over its enemies. Its ground force in South Viet-Nam was far too small by traditional standards (by a factor of between three and ten) to attain such an end there. Indeed, when compared on a monthly basis, US battle deaths were twice as high during its Korean War as during the Second Indochina War, and a full 15 times as high during World War II (Associated Press, 1975). The United States attempted to compensate for this deficit in military manpower and to

Table 1.4. Some measures of US effort during the Second Indochina War: a breakdown by region

Region	On an area basis[a]		On a population basis[a]	
	Munition expenditure[b] *kg/ha*	Herbicide expenditure[c] *l/ha*	Munition expenditure[b] *kg/capita*	Herbicide expenditure[c] *l/capita*
South Viet-Nam	587	4.2	577	4.1
Military Region I	1 166	4.4	1 066	4.0
Military Region II	268	2.0	669	4.9
Military Region III	1 431	12.7	890	7.9
Without Saigon			1 833	16.3
Military Region IV	134	1.7	77	1.0
North Viet-Nam	67	. .	57	. .
Cambodia	42	. .	113	. .
Laos	94	. .	773	. .
Total	**189**	**1.0**	**306**	**1.6**

Notes and sources:

[a] Area and population data are from table 1.1.

[b] Munition data are from table 2.2.

[c] Herbicide data are from table 3.5. To convert any of these volume data to average kilograms of active ingredients, multiply by 0.756 918.

tilt the military balance in its favour by a variety of means. These included occasional punitive ground raids (the so-called search-and-destroy missions), carried out most often in South Viet-Nam, sometimes in eastern Cambodia, and rarely in southeastern Laos. They further included the employment of technologically sophisticated weapons and techniques and the lavish expenditure of remotely delivered munitions (Hymoff, 1971; White, 1974:II).

Important in the present context among the several interrelated cost-intensive rather than manpower-intensive means by which the USA attempted to subdue its guerrilla enemy in South Viet-Nam were: (*a*) local forest destruction (primarily to deny the enemy freedom of movement, staging areas and cover in general); (*b*) local crop destruction (primarily to deny the enemy local sources of food and other resources); (*c*) forced relocation of indigenous civilians into the US controlled areas (primarily to deny the enemy local logistical and other support); and (*d*) disruption of the supply lines from the three surrounding countries (again primarily to deny the enemy logistical, manpower and other support).

IV. *Conclusion*

The Second Indochina War was a local war fought with conventional weapons. It was, however, an innovative war in that a great power attempted to subdue a peasant army through the profligate use of technologically advanced weapons and techniques. A number of these weapons and techniques were inescapably anti-ecological. It is these anti-ecological tactics which are the major focus of the present monograph. Of particular importance in this context are the following novel approaches to widespread and long-term rural area denial: the massive employment of high-explosive munitions (chapter 2), the lavish application of anti-plant chemicals (chapter 3), and the large-scale use of heavy landclearing tractors (chapter 4).

Notes

[1] The term "Second Indochina War" refers to the hostilities between the United States on the one hand and South Viet-Nam, North Viet-Nam, Cambodia and Laos on the other during the period 1961–73. The French hostilities in the region during the period 1946–54 are, by implication, considered to comprise the "First Indochina War". In the United States the Second Indochina War is referred to as the "Vietnam Conflict".

[2] The term "ecocide" is meant to signify the large-scale destruction of wild plants and animals and their natural habitat, that is, the destruction of natural ecosystems. The term appears to have been coined by A.W. Galston (Times, 1970) and soon came into rather widespread use (for example, Neilands, 1970a; Weisberg, 1970; Johnstone, 1970–1971; Westing, 1971a). R.A. Falk has drawn the parallel between "ecocide" and "genocide" and proposes an international convention against the crime of ecocide coordinate in a sense with the one against the crime of genocide (*see* page 86).

[3] There is a voluminous literature on Indochina. The US Department of Defense area handbooks for the region provide very useful basic sources of information. *See* those for South Viet-Nam (H. Smith *et al.*, 1967b), North Viet-Nam (H. Smith *et al.*, 1967a), Cambodia (Whitaker *et al.*, 1973) and Laos (Whitaker *et al.*, 1972). A map (National Geographic Society, 1967) and a number of atlases and geographies are also available (Engineer Agency . . . , 1968; Frodigh *et al.*, 1969; Matheson & Laurendeau, 1953; Stuttard *et al.*, 1943). For information on the climate, *see* Engineer Agency . . . (1968), Grimes (1967), H. Hodges & Webb (1968), Nuttonson (1963), and Ohman (1965). For information on the soil, *see* Moormann (1961) and on the vegetation, *see* L. Williams (1965; 1967), Tung (1967), and Oakes (1967).

Among the bibliographies dealing with Indochina can be mentioned those by Anglemyer *et al.* (1969), US Army (1972), Carlson (1974), Fisher (1967), Gee & Anglemyer (1970), Grimes (1967), C. Hobbs (1964), C. Hobbs *et al.* (1950), Leung *et al.* (1972), Mekong Documentation Centre (1967), State (1963a; 1963b), and Tregonning (1969).

[4] The literature dealing with the Second Indochina War is overwhelming not only in terms of its magnitude, but also regarding such questions as scholarliness, objectivity and so on. FitzGerald (1972) provides an excellent historical background and Fulbright (1970) a useful collection of documents. Four US government publications ought to be mentioned: two in the open literature, by Sharp & Westmoreland (undated) and by the US Marine Corps (1974), and two analyses of the war which were at one time secret, the so-called Pentagon Papers (Defense, 1968; *see also* Gravel *et al.*, 1971–1972) and National Security Study Memorandum No. 1 (Dellums, 1972). A number of relevant bibliographies are available (Leitenberg & Burns, 1973; Westing, 1974a). Finally, the US Department of the Army in 1973 began publishing a series of monographs under the general heading, "Vietnam Studies", projected to include about 22 volumes (*see, for example*, Hay, 1974; Ploger, 1974; Rogers, 1974).

Chapter 2. High-explosive munitions

Where indicated thus,[1] the reader is referred to the notes on page 22.

I. *Introduction*

Bombing and shelling with high-explosive munitions have been inseparable parts of modern warfare.[1] Indeed, bombing and shelling provided the mainstay of the US effort during the Second Indochina War. During this war such conventional munitions were delivered in quantities unprecedented in military history. Although their environmental impact seems to have been generally unrecognized, they have left an indelible mark on the land and people of the region.

Not only were the total munition expenditures in Indochina extraordinarily high, but the overall fraction of them devoted to ill-defined rural area targets was also of unprecedented magnitude.

The present chapter is devoted first to a description of the employment of these conventional high-explosive munitions in Indochina (section II) and then to a consideration of their specific ecological impact (section III). The more general and long-range ecological aspects of such actions are covered elsewhere (*see* chapter 6), as are certain military considerations (*see* page 83).

II. *Description*

During the eight and a half years for which figures have been released, the USA expended a total of over 14 x 10^6 tonnes of munitions against Indochina. More than one-third of this amount was delivered during the two peak years 1968 and 1969. Just half was delivered from the air and the remainder from surface weaponry (table 2.1).

The Second Indochina War was far from being spread evenly over the face of Indochina. For a better understanding of its environmental impact it thus becomes necessary to examine the overall munition expenditures on a regional basis. When this is done it becomes evident at once that South Viet-Nam was by far the most severely affected region. Representing less than a quarter of Indochina's land area, South Viet-Nam nonetheless was the recipient through the years of 71 per cent of all the munitions expended by the USA (table 2.2). In order to take this analysis one step further it becomes necessary, of course, to consider these regional munition expenditures taking into account the differing sizes of each of the regions, and perhaps also their differing population densities (table 2.3). This points out even more dramatically the disproportionate impact of the war on the land and people of South Viet-Nam – and in particular, on South Viet-Nam's Military Regions III and I.

Table 2.1. US munition expenditures in the Second Indochina War: a breakdown by year and major delivery system

10^6 kg = 10^3 tonnes

Year	Air			Ground	Sea	**Total**
	Total air	F-Bs etc.	B-52s			
1961	. .	. .	–	. .	. .	. .
1962	. .	. .	–	. .	. .	. .
1963	. .	. .	–	. .	. .	. .
1964	. .	. .	–	. .	. .	. .
1965	286	257	29	. .	. .	**286**
1966	458	365	93	535	5	**998**
1967	846	609	237	1 092	27	**1 965**
1968	1 304	772	531	1 347	46	**2 696**
1969	1 258	758	501	1 275	27	**2 561**
1970	887	516	370	1 072	12	**1 970**
1971	692	397	295	756	5	**1 453**
1972	984	485	499	776	32	**1 792**
1973	378	138	241	163	3	**543**
Total	**7 093**	**4 297**	**2 796**	**7 016**	**156**	**14 265**

Notes and sources:

(*a*) The total-air datum for 1965 is from the US Library of Congress (1971:8). The total-air and ground data for 1966–73 are from a US Department of Defense release of 17 August 1973. The sea data for 1966–70 are from the US Library of Congress (1971:8) whereas those for 1971–73 are from the US Department of Defense (private letter, 2 November 1973).

(*b*) B-52 data are from a US Department of Defense release of 18 July 1973 (with appropriate extrapolation for the missing four months). The air munitions other than B-52 (fighter-bombers, etc.) are the difference between the total-air data and the B-52 data.

(*c*) The 1973 air data are for January–July. The 1973 ground data are for January–May.

(*d*) Missing data have not been released by the US Department of Defense.

During most of the war much of rural South Viet-Nam and of the rest of Indochina were designated as free-fire zones or euphemistically as "specified-strike" zones (*see* Goldwater, 1975). The US Department of Defense has released no information on the extent of the free-fire zones in the various countries and none on how the sizes of these zones have varied with time. It has been reported that the free-fire zones in Military Region I of South Viet-Nam covered over 90 per cent of its area (Kennedy, 1971:III:25). A reasonable minimum estimate for the whole of Indochina would be that the free-fire zones covered three-quarters of the entire region. In the light of this it should come as no surprise that no habitat in Indochina was spared the impact of high-explosive munitions, including forests and swamps, fields and paddies (*see* pages 1–6).

Table 2.2. US munition expenditures[a] in the Second Indochina War: a breakdown by region and major delivery system

10^6 kg = 10^3 tonnes

Region	Air[b]					
	Total air	F-Bs etc.	B-52s	Ground[c]	Sea[d]	**Total**
South Viet-Nam[e]	3 285	1 798	1 487	6 875	16	**10 176**
Military Region I	1 059	579	479	2 215	5	**3 279**
Military Region II	667	365	302	1 396	3	**2 066**
Military Region III	1 395	764	632	2 920	7	**4 322**
Military Region IV	165	90	75	345	1	**510**
North Viet-Nam	961	802	159	–	141	**1 101**
Cambodia	614	275	339	140	–	**754**
Laos[f]	2 233	1 422	811	–	–	**2 233**
Total	**7 093**	**4 297**	**2 796**	**7 016**	**156**	**14 265**

Notes and sources:

[a] Basic munition data are from table 2.1.

[b] Air munitions are partitioned among the countries in accordance with the proportions released by the US Department of Defense on 18 July 1973 (with the missing final four months partitioned 90 per cent to Cambodia and 10 per cent to Laos).

[c] Ground munitions are partitioned 98 per cent to South Viet-Nam and 2 per cent to Cambodia.

[d] Sea munitions are partitioned 90 per cent to North Viet-Nam and 10 per cent to South Viet-Nam.

[e] All munitions expended in South Viet-Nam are partitioned among the Military Regions in accordance with the proportions of B-52 missions flown against these regions during the four years 1967–70 (Littauer & Uphoff, 1972:277–78).

[f] Although no breakdown is provided in the above table for the Laotian data, it has been reported that during 1965–71 70 per cent of the total was expended in southern Laos against the so-called Ho Chi Minh Trail, and the remainder in the rest of the country (Littauer & Uphoff, 1972:281).

Also crucial to an understanding of the environmental impact of the high-explosive munitions employed in such massive quantities in Indochina is a recognition of the strategy behind their employment (*see* page 6) and how this influenced the nature of their targeting. It turns out that an extraordinarily high proportion of the bombs and shells expended in Indochina were directed against ill-defined rural targets. Indeed, it appears that of the order of 90 per cent of all munitions expended by the United States during the Second Indochina War was in the form of unobserved fires aimed at rural area targets, often nebulously defined as harbouring "suspected enemy activity" (Dellums, 1972:16754; Enthoven & Smith, 1971:305; Kennedy, 1967:145; Littauer & Uphoff, 1972:53). It is possible to see occasional, scattered craters almost anywhere in Indochina and regions of intense craterization in many localities, including especially War Zones C and D, the Iron Triangle, the Rung Sat and

Table 2.3. US munition expenditures[a] in the Second Indochina War: a breakdown by region, per unit area and per capita

Region	Munitions per area[b] *kg/ha*	Munitions per capita[b] *kg/capita*
South Viet-Nam	587	577
Military Region I	1 166	1 066
Military Region II	268	669
Military Region III	1 431	890
Without Saigon		1 833
Military Region IV	134	77
North Viet-Nam	67	57
Cambodia	42	113
Laos	94	773
Total	**189**	**306**

Sources:

[a] Basic munition data are from table 2.2.

[b] Areas and populations are from table 1.1.

U Minh forests, the Fish Hook and Parrot's Beak regions of Cambodia, the Demilitarized Zone at the border between North and South Viet-Nam, and the tri-border region where Laos, Cambodia and South Viet-Nam all meet. A US Air Force officer has described that entire region in southern Laos referred to as the Ho Chi Minh Trail as being cratered "like the backside of the moon" (Kamm, 1971:4).

The employment of B-52 Stratofortresses for the massive delivery of high-explosive munitions to area targets deserves special attention in the present context. Indeed, this innovative approach to counterinsurgency warfare seems to have been highly important to the US effort in this regard.[2] These immense bombers were modified for the delivery of conventional bombs and pressed into Indochina service in mid-1965. During the ensuing eight years they were used on a daily basis.

The importance of the B-52s to the US war effort is well demonstrated by their high level of employment: the 2.8 x 10^6 tonnes of high-explosive munitions delivered by them amounted to 39 per cent of all air munitions, or to 20 per cent of the total of all munitions, expended during the war (table 2.1). They were particularly important in Cambodia and Laos, where this highly impersonal, relatively inaccurate, and rather indiscriminate form of delivery accounted for 45 and 36 per cent respectively, of total US munition expenditures in those countries (table 2.2).

The maximum load per B-52 aircraft was 108 of the "500-pound" (241-kg) bombs (the most usual munition) or 66 of the "750-pound" (362-kg) bombs (Drendel, 1968:48–50). The actual load delivered by these aircraft each time they went on a bombing mission in Indochina throughout their eight years of service came to an average of 22.5 tonnes (87 per cent of the potential maximum). The average wartime load thus consisted of just over 93 "500-pound" bombs (or their equivalent) per sortie (that is, per aircraft). The B-52s averaged 42 sorties per day throughout the war. Each day, therefore, the B-52s alone accounted for the release (on the average) of another 3 891 such bombs somewhere over Indochina.

The B-52s flew their counterinsurgency missions from an altitude of 10 km and more in groups that averaged 4.6 in number, dropping their bomb loads through the usual cloud cover with the aid of an onboard electronic computer. Lack of pinpoint accuracy was compensated for by the carpet-bombing approach employed. The bomb delivery system is designed so that all of the bombs released from one aircraft form a saturation pattern on the ground which is presumably meant to produce casualties throughout this target area. This so-called target box is roughly in the form of an elongated rectangle and has an area of approximately 65 ha (Littauer & Uphoff, 1972:25; Whitney, 1972).

With a strike zone of 65 ha per B-52 aircraft and an average of 4.6 aircraft per Indochina mission, the average zone of devastation per B-52 mission can be seen to have been 296 ha. Inasmuch as an average of nine such missions were flown daily throughout the war, the total area carpeted by the B-52s at one time or another was 8.1×10^6 ha – just over 11 per cent of the total area of Indochina. If one singles out South Viet-Nam (a region accounting for 53 per cent of the B-52 bomb tonnage expenditures), the comparable data are even more impressive: here the target boxes add up to 4.5×10^6 ha, in other words, to fully 26 per cent of South Viet-Nam's area. These values are, of course, inflated to the extent to which any particular area was rebombed. Since a major fraction of the B-52 missions was for continuing harassment and area denial in regions controlled by the US enemy through the years, there must have been a considerable amount of target overlap, but how much is not readily available.

Finally in this section it will be useful to summarize what is known regarding the high-explosive munitions that were employed in Indochina. Of particular interest here are the major types of environmentally disruptive munitions used and the frequency of their employment. The most commonly used bomb was the "500-pound" (241-kg) bomb, whereas the most commonly used artillery shell was the 105-mm howitzer projectile. These and several of the other more frequently used munitions are described in table 2.4. Although the numerical rankings of these munitions are available, their proportionate frequencies of use are not. Equally unfortunate is the fact that no information has been released on the relative frequencies of use of the various fuses. These are both important considerations with respect to environmental impact, making it

Table 2.4. Munitions most commonly employed by the USA during the Second Indochina War[a]

Service	Numerical ranking	Munition[b]	Actual weight[c] *kg*	Weight of explosive *kg*
Air Force	1	"500-pound" bomb (MK-82)[d]	240.9	87.1
	2	"750-pound" bomb (MK-117)	362.4	175.1
	3	"2 000-pound" bomb (MK-84)	893.6	428.6
Army	1	105-mm howitzer projectile[e]	13.0	2.3
	2	155-mm howitzer projectile	43.4	7.0
	3	8-inch (203-mm) howitzer projectile	91.2	. .
	4	175-mm gun projectile	66.7	. .
Navy	1	5-inch (127-mm) gun projectile	24.7	(3.3)

Notes and sources:

[a] The information provided in the table above was supplied by several services: US Air Force (private letters, 13 November 1973, 11 January 1974), US Army (private letters, 14 November 1973, 28 March 1974), and US Navy (private letter, 8 November 1973).

[b] Air-delivered munitions accounted for 50 per cent of total expenditures in Indochina, ground-delivered for 49 per cent, and sea-delivered for 1 per cent (table 2.1).

[c] The weights given in pounds are only nominal weights whereas those given in kilograms are actual weights.

[d] The MK-82 "500-pound" general-purpose (GP) bomb, the most frequently used bomb during the Second Indochina War, has the following characteristics (Army, Navy & Air Force, 1966; US Air Force, private letters, 13 November 1973, 13 January 1974, 2 October 1975, 17 November 1975): total weight 240.9 kg, weight of explosive 87.1 kg, assembled length 220.7 cm, maximum body diameter 27.3 cm, fin span 38.4 cm. It should be noted that the final actual "drop" (or "all-up") weight will be more than the 240.9 kg given above, depending upon the attachments. Thus, when the bomb is outfitted in the "conical-fin non-retard" configuration the drop weight is 261.7 kg, and when outfitted in the "retard" mode it is 287.1 kg. Explosive filler: "tritonal", a 4:1 mixture (by weight) of 2,4,6-trinitrotoluene (TNT) and powdered aluminium, or else "H6", a 9:6:4:1 mixture (by weight) of hexahydro-1,3,5-trinitro-*s*-triazine ("cyclonite", "RDX"), TNT, powdered aluminium, and a derivative of ammonium picrate known as "D2", to which is added a trace (0.5 per cent) of calcium chloride. Blast (shock) wave peak overpressures (French & Callender, 1962:106): 552 kPa at 9.1 m, 41 kPa at 18.3 m, 21 kPa at 27.4 m. Initial fragment velocity (French & Callender, 1962:113): 2 252 m/s. (Note that the blast wave and fragment velocity data may be somewhat out of date.)

[e] The M-1 high-explosive (HE) 105-mm howitzer cartridge, the most frequently used artillery shell during the Second Indochina War, has the following characteristics (Army, 1967a:II:113–15): weight of complete round 19.1 kg, weight of projectile alone 13.0 kg, weight of explosive in projectile 2.3 kg, length of complete round 78.9 cm, length of projectile alone 49.4 cm, diameter of projectile 105 mm. Explosive filler in projectile: "Composition B", a 3:2 mixture (by weight) of hexahydro-1,3,5-trinitro-*s*-triazine ("cyclonite", "RDX") and 2,4,6-trinitrotoluene (TNT), to which may be added a small amount (1 per cent) of wax. Ballistics: Maximum muzzle velocity 472 m/s, maximum range 11 275 m.

necessary in the next section to make some estimations regarding these parameters.

III. *Ecological consequences*

The environmental impact of high-explosive munitions is divisible more or less easily into the immediate or early effects, considered here, and the long-term effects, considered elsewhere (chapter 6). Among the early effects are the consequences of the initial explosion and of the resulting crater. The former include the effects of the blast and of the flying metal and the latter of local destruction of the vegetative cover, displacement and exposure of earth, and possible disruption of the water table and of local drainage patterns. Important effects in the months and even years following the attack include accelerated soil erosion (page 64), loss of nutrients in solution or nutrient dumping (page 66), and changes in the biotic community, both of the plants (page 69) and of the animals (page 71).

In Indochina the habitat disruption brought about by the massive bombing and shelling was in places truly spectacular. As indicated earlier, aerial views of crater fields have evoked a moonscape analogy from military and non-military observers alike. Military sources can be quoted for ground-level impressions as well: "the landscape [was] torn as if by an angry giant. The bombs uprooted trees and scattered them in crazy angles over the ground. The tangled jungle undergrowth was swept aside around the bomb craters" (R. Kipp, 1967–1968:17). In another instance, an infantry officer explains that "fighter-bombers and artillery pound the [land] into the gray porridge that the green delta land becomes when pulverized by high explosives" (Arnett, 1969).

There is, of course, no denying that bombing and shelling destroy fauna and flora in the target area and otherwise disrupt the habitat there. The next question is thus how extensive this damage was in Indochina. Unfortunately not all of the necessary information is available for such a determination. Thus, providing a preliminary answer to this question requires that the calculations be based on a number of more or less well-informed assumptions. As a prerequisite to all further steps in this process it becomes necessary to estimate what fraction of the total munitions expended (a value that has been released, *see* table 2.1) was of the sort that produces craters. This crater-producing fraction is estimated here to be 10.6 x 10^6 tonnes, or 74 per cent of total munition expenditures in Indochina (table 2.5).

The first item of potential environmental concern is the initial blast (shock) wave generated by the explosion. In fact, the peak transient overpressure of the outwardly moving shock front diminishes to insignificance before it travels very far. Thus, the lethal area of a "500-pound" (241-kg) bomb based on the blast overpressure it generates is significantly less than 10 metres in diameter (that is, less than 0.01 ha in area) (French & Callender, 1962:105–106; War,

Table 2.5. Environmental impact of US munition[a] expenditures during the Second Indochina War

Region	Crater-producing munitions[b] *10⁶ kg*	Area of flying "shrapnel"[c] *10³ ha*	Surface area of craters[d] *10³ ha*	Volume of craters *10⁶ m³*
South Viet-Nam	7 067	8 797	147	1 964
Military Region I	2 278	2 834	47	633
Military Region II	1 434	1 786	30	399
Military Region III	3 001	3 736	62	834
Military Region IV	354	441	7	98
North Viet-Nam	941	1 171	20	262
Cambodia	643	800	13	179
Laos	1 949	2 426	40	542
Total[e]	**10 599**	**13 194**	**220**	**2 947**

Notes and sources:

[a] Basic munition data are from table 2.2.

[b] Amounts of crater-producing munitions are obtained additively, using 80 per cent of the amounts delivered by fighter-bombers and other aircraft, 100 per cent of those by B-52s, 60 per cent of those by ground-based delivery systems, and 100 per cent of those by sea-based delivery systems (estimated percentages in the absence of released US Department of Defense information). (A weighted average of the air-delivery systems would be 88 per cent, and one for all of the means combined would be 74 per cent.)

[c] Area of flying metal shards or "shrapnel" is obtained by assuming that all of the crater-producing munitions had been delivered in the form of "500-pound" (241-kg) bombs and that each of these had scattered significant amounts of "shrapnel" over an area of 0.3 ha.[3]

[d] Surface area of craters is based on an average crater diameter of 8 m, and thus an area of 0.005 ha. Their volume is based on an average maximum depth of 4 m, and thus a volume of 67 m^3.[4]

[e] The data presented do not consider overlap, information that is not readily available. Tentative values might be 25 per cent for area of flying "shrapnel" and 5 per cent each for surface area and volume of craters.

1943:23). This is substantially smaller than the lethal area (by perhaps an order of magnitude) resulting from the fragmentation effect, discussed below.

The zone of fragmentation or flying metal that results from the use of high-explosive munitions appears to be a highly significant aspect of the problem. In Indochina this zone of flying metal (or "shrapnel", as it is commonly called) added up to some 13 x 10^6 ha, that is, to 17 per cent of the total Indochina region (table 2.5). Moreover, if South Viet-Nam alone is considered in this regard, the comparable area was 9 x 10^6 ha, representing more than 51 per cent of that country, not considering overlap.

Clearly some fraction of the wildlife of Indochina was killed or wounded by "shrapnel". On the other hand, it is difficult to estimate either the amount of wildlife thus killed or the ecological impact of this mortality. Similarly, millions of trees must have been killed by blast or "shrapnel", either directly or indirectly. If one considers merely the trees that had been growing on land today occupied by craters – a combined area of more than 200×10^3 ha throughout Indochina (table 2.5) – and estimates that all of this cratered land had supported, on the average, 10 trees per hectare having a diameter at breast height (or DBH) of 10 cm or greater, then one can assume an immediate destruction of at least 2×10^6 such trees. To this amount must next be added the many trees in the 0.3 ha zone directly surrounding each bomb crater which were knocked down by the blast, cut down by the flying fragments, or which died subsequently as an indirect consequence of "shrapnel" wounds in their trunks. Assuming that one-third of such trees succumb, this would bring the total number of trees killed to more than 45×10^6 (*see also* page 70).

"Shrapnel" wounds can result in the delayed death of a tree by providing a site of entry for wood-rotting fungi. The spreading rot can weaken the bole to the point where the wind breaks it off within several years. Species of *Dipterocarpus* and *Anisoptera* (both Dipterocarpaceae) as well as *Hevea brasiliensis* (Euphorbiaceae) are quite rapidly vulnerable to destruction of this sort whereas species of *Hopea* (Dipterocarpaceae) and *Lagerstroemia* (Lythraceae) are more resistant (Westing, 1972b). It has been estimated on the basis of logs coming into local sawmills that over four-fifths of the trees in South Viet-Nam's War Zone D were struck by flying metal.

Leaving aside the effects of initial blast and fragmentation, next to be considered are the effects of the residual craters. The nature and severity of their effect hinge upon a number of factors, important among them the density and extent of the crater field, the character of the original vegetative cover, the character of the rainfall and other aspects of the local climate, the slope and other topographic features, the soil type and the type of and depth to the water table.

Transpiration will be reduced (and thus evaporative water loss or flyoff) to the extent that the vegetation has been destroyed and for the period of months (or, in some situations, years) before a comparable vegetational cover has become re-established (*see* page 67). During this period the water table will rise to a higher level. This raised water table may in turn exert an adverse effect on any extant vegetation that had been spared direct battlefield damage. Moreover, during the time that flyoff is reduced there is likely to be a greater streamflow out of the affected area, and thus perhaps an aggravation of downstream flood danger. The increased runoff will result in a higher level of erosion (particulate loss) and of nutrient loss in solution to the point of so-called nutrient dumping. The soil which had been scattered by the explosion is particularly subject to being carried off, in the runoff during wet periods and by the wind during dry periods.

If the original water table had been close enough to the surface for the craters to penetrate it, then these will fill with water. The subsequent evaporation from the open water surfaces is, however, likely to be roughly comparable in magnitude to the transpiration from the vegetation which had previously occupied the cratered locations, and thus no important change in flyoff occurs from such sites.

Should the craters have breached a hardpan, the perched water table would be lost for many years. The resulting change in site conditions would be likely to exert a significant effect on the composition of the vegetative cover.

A huge amount of soil (both topsoil and subsoil) was displaced by the high-explosive munitions expended in Indochina. A rough calculation suggests this amount to have been of the order of 3×10^9 m^3 (table 2.5). Although some fraction of this was thrown out of the craters – some to form a lip and some more widely dispersed – the bulk of it seems to have been compacted into the sides of the crater. Based on experience with the civil use of dynamite for ditching, it appears that the soil compaction in the sides of the craters is likely to diminish to insignificance through the action of natural forces over a period of several years (J.H. Patric, US Forest Service, private letter, 25 February 1974).

In considering the environmental impact of craterization there is, of course, the important question of a crater's longevity. Indeed, under ordinary field conditions the longevity of a crater appears to be many years or decades. Thousands of craters can be found in Indochina today with consistently less than a metre of soil washed back into them, particularly in relatively flat terrain. Within four or five years the sides of those craters that do not become ponds become stabilized by vegetation. They are then a seemingly permanent feature of the landscape. For example, the shell craters produced during the battle of Verdun in 1914 and during other major World War I battles were found to be still present and only sparsely vegetated some five decades after they were produced (Life, 1964; Westing & Pfeiffer, 1972).

With the longevity of craters in mind, it becomes of interest to estimate their numbers in Indochina. On the basis of the relative frequencies of the various crater-producing munitions expended (table 2.4) and considering their combined weight (table 2.5), it becomes possible to calculate that roughly 21×10^6 bombs and 229×10^6 shells were expended there – and thus, that there now exist almost the same number of permanent craters, large and small.

IV. *Conclusion*

The present chapter suggests that a new era in warfare has been ushered in with the *en masse* employment in Indochina of high-explosive munitions, expended largely for the purpose of long-term and widespread rural area denial. This tactical innovation has led to such a high level of indiscriminate destruction of field and forest that it must be evaluated not only in the traditional terms of human impact (both military and civil), but also in terms of long-term site or habitat disruption. Such newly imposed ecological considerations thus add a significant temporal dimension to the consequences of war. One can only hope that the military failure of this attempt at "massive strategic interdiction" in Indochina will discourage the adoption of a similar strategy by some belligerent in a future counterinsurgency or other war (*see* pages 83–85).

Notes

[1] For descriptive and historical literature dealing with artillery, *see* Hogg (1970), Manucy (1949), Rogers (1971), and Stevens (1965). An analysis of the World War II bombing effort by the Allies has been made by the US Strategic Bombing Survey (D'Olier *et al.*, 1947). The air war in Korea has been covered by Futrell *et al.* (1961) and that of the Second Indochina War by Harvey (1967) and by Littauer & Uphoff (1972).

The environmental impact of high-explosive munitions has been pointed out a number of times in recent years (Pfeiffer, 1969–1970; 1971; Lewallen, 1971:IV; Westing & Pfeiffer, 1972; Lang *et al.*, 1974:IV:51, 86–87).

[2] The early history of the B-52 "Stratofortress" bomber aircraft (Boeing Co., Seattle, Washington, USA) in Indochina has been reviewed by R. Kipp (1967–1968). A wealth of statistical information on the use of the B-52s in Indochina was released by the US Department of Defense on 18 July 1973. The numerical information presented here on their use is derived largely from this release, with modest supplementation from the data of Littauer & Uphoff (1972:277–78).

[3] Estimating the average size of zone within which a high-explosive munition creates environmentally significant damage via blast and fragmentation is both difficult and arbitrary. Important variables include the type of munition (for example, bomb *versus* shell, "general purpose" *versus* "fragmentation"), the size or weight of munition (and of its explosive content), the type of fuse (for example, quick *versus* delay), the character of the terrain and vegetative cover, and, not least, what is meant by the appellation "environmentally significant".

The estimate used here of 12.5 m^2/kg of high-explosive munition is meant to be an integration of all these considerations in order to provide a conservative overall average value for Indochina. Translating the estimate used into its equivalent for a "500-pound" (241-kg) bomb (the most frequently used high-explosive air munition, accounting for perhaps 30 per cent by weight of the total of all munitions expended by the USA during the Second Indochina War), the area of damage becomes 0.30 ha. Doing the same for a 105-mm (13-kg) howitzer shell (the most frequently used high-explosive ground munition, and accounting for perhaps another 30 per cent of the overall total), the area becomes 0.016 ha.

By way of comparison, a "500-pound" bomb is said to be "lethal" for a person standing in the open (Stillman, 1972) and the woody vegetation effectively demolished (Lang *et al.*, 1974:IV:51) within an area of 0.29 ha. Moreover, a bursting, quick-fused, high-explosive 105-mm howitzer shell produces a zone 0.049 ha in size in which a person standing is known to have a 50 per cent chance of becoming a casualty (Army, 1950:19), suggesting the above-used value to be quite a conservative one as an overall average.

[4] The size of the crater and the amount of soil displacement brought about by high-explosive munitions depend not only on the size of the munition used, but also on the type of fuse employed, the character of the ground, and other factors. An estimated 85 to 95 per cent of all general-purpose bombs dropped in Indochina were set for some delay (US Air Force, private letter, 2 October 1975). The dimensions used here as overall Indochina averages are derived from an estimate of the average size of a "500-pound" (241-kg) bomb crater. The diameter of such a crater has been assumed to be 8 m and its maximum depth 4 m. These values lead to a surface area or opening of 0.005 ha (that is, 0.209 m^2/kg) and, assuming a conical shape, a volume of soil displacement of 67 m^3 (that is, 0.278 m^3/kg).

By way of comparison, the average diameter of a "500-pound" (241-kg) bomb crater in Indochina has been reported to be 9 m (Littauer & Uphoff, 1972:222) and even 10 m (Lang *et al.*, 1974:IV:51). A 250-kg bomb with delay fuse and containing 56.7 kg of explosive (that is, 35 per cent less than the current "500-pound" model) was found to have an average crater diameter of between 5.5 m and 7.6 m (War, 1943:29). Finally, a 16-inch (406-mm, 862-kg) naval gun shell has been reported to produce a crater with a diameter of 9 m (Whitney, 1969).

Chapter 3. Anti-plant chemicals

Where indicated thus,[1] the reader is referred to the notes on page 41.

I. *Introduction*

Man, in common with all animals, is dependent upon the food and shelter he derives from the plant kingdom. The intentional military destruction during war of vegetation in territory under enemy control is a form of recognition of this fundamental relationship. Indeed, crop destruction has been a continuing part of warfare for millennia and the military importance of forests has also long been recognized.

The present chapter deals with a modern approach to this intentional vegetational destruction, that is, the employment of plant-killing chemicals or herbicides for military purposes.[1] A description of their use during the Second Indochina War (section II) is followed by the ecological consequences peculiar to such an assault on the environment (section III). The more general and long-range ecological aspects of such actions are covered elsewhere (chapter 6), as are certain military considerations (*see* page 83).

Although herbicides are widely used in agriculture, forestry, and other fields, it is of interest to note that their development is closely linked with chemical warfare research during the 1940s (Peterson, 1967). The first military use of modern chemical anti-plant agents appears to have occurred on a modest scale in Malaya during the 1950s (Henderson, 1955), an action totally dwarfed by that in Indochina during the subsequent decade. Indeed, the military use of anti-plant agents during the Second Indochina War stands out as one of the most controversial aspects of that conflict.

II. *Description*

The United States carried out a massive herbicidal programme during the Second Indochina War that stretched out over a period of almost a decade. This programme was aimed for the most part at the forests of South Viet-Nam and to a lesser extent at its crops (Westing, 1972d). Using a variety of agents, the USA eventually expended a volume of more than $72 \times 10^3 m^3$ containing almost 55×10^6 kg of active herbicidal ingredients.

The major anti-plant agents that were employed by the United States in Indochina – colour-coded "Orange", "White", and "Blue" – are described in some detail in table 3.1. Agents Orange and White consist of mixtures of plant-hormone-mimicking compounds which kill by interfering with the normal metabolism of poisoned plants. Agent Blue, on the other hand, consists of a desiccating compound which kills by preventing a plant from retaining its moisture content.[1] Agents Orange and White are particularly suitable for use against dicotyledonous plants, whereas Agent Blue is relatively more suitable for use against monocotyledonous plants. At the high levels used for military application (28 litres/ha, averaging 21 kg/ha in terms of active ingredients) these herbicides are, however, not so selective as one might expect on the basis of civil experience.[2]

Table 3.1. Major anti-plant agents employed by the USA in the Second Indochina War

Type	Composition	Physical properties	Application
Agent Orange	A 1.124:1 mixture (by weight) of the n-butyl esters of 2,4,5-trichlorophenoxyacetic acid (2,4,5-T) (545.4 kg/m^3 acid equivalent) and 2,4-dichlorophenoxyacetic acid (2,4-D) (485.1 kg/m^3 acid equivalent)	Liquid, oil soluble, water insoluble, weight 1 285 kg/m^3	Applied undiluted at 28.062 litres/ha, thereby supplying 15.306 kg/ha of 2,4,5-T and 13.612 kg/ha of 2,4-D, in terms of acid equivalent, and also an estimated 70.155 mg/ha of the contaminant dioxin (2,3,7,8-tetrachlorodibenzo-*p*-dioxin)
Agent White	A 3.882:1 mixture (by weight) of the tri-isopropanolamine salts of 2,4-dichlorophenoxyacetic acid (2,4-D) (239.7 kg/m^3 acid equivalent) and 4-amino-3,5,6-trichloropicolinic acid (picloram, "Tordon") (64.7 kg/m^3 acid equivalent)	Aqueous solution, oil insoluble, weight 1 150 kg/m^3	Applied undiluted at 28.062 litres/ha, thereby supplying 6.725 kg/ha of 2,4-D and 1.816 kg/ha of picloram, in terms of acid equivalent
Agent Blue	A 2.663:1 mixture (by weight) of Na dimethyl arsenate (Na cacodylate) and dimethyl arsinic (cacodylic) acid (together 371.460 kg/m^3 acid equivalent)	Aqueous solution, oil insoluble, weight 1 310 kg/m^3	Applied undiluted at 28.062 litres/ha, thereby supplying 10.424 kg/ha, in terms of acid equivalent (and 5.659 kg/ha of elemental As)

Notes and sources:

(*a*) The above information is derived primarily from Darrow *et al.* (1969:IV), and secondarily from Irish *et al.* (1969) and Young & Wolverton (1970). *See also* Young (1974:I).

(*b*) For the basis of the dioxin concentration in Agent Orange, *see* note 9, page 44.

(*c*) Numerous herbicidal formulations have been tested by the USA as chemical anti-plant agents, several of which were assigned a colour code during their more or less ephemeral existence: "Orange II" was similar to the "Orange" above except that its 2,4,5-T moiety was replaced by the iso-octyl ester of 2,4,5-T. "Pink" was a mixture of the n-butyl and iso-butyl esters of 2,4,5-T. "Green" consisted entirely of the n-butyl ester of 2,4,5-T. And "Purple" was a mixture of the n-butyl ester of 2,4-D and the n-butyl and iso-butyl esters of 2,4,5-T.

Of the several agents used, it appears that Orange was the agent of choice since it represented 61 per cent of the total volume expended over the years (table 3.2). In each of the three peak years of spraying – 1967 through 1969 – about equal magnitudes of these agents were expended, together accounting for over three-quarters of the total US wartime expenditures. These were also very active war years in other respects (table 1.3).

Forest destruction was generally accomplished through the use of Agents Orange or White, whereas Agent Blue was usually the agent of choice for the destruction of rice (*Oryza sativa*, Gramineae) and other crops, although Agent

Table 3.2. US herbicide expenditures in the Second Indochina War: a breakdown by agent and year

$m^3 = 10^3$ litres

Year	Agent Orange	Agent White	Agent Blue	**Total**
1961	. .	–	. .	. .
1962	56	–	8	**65**
1963	281	–	3	**283**
1964	948	–	118	**1 066**
1965	1 767	–	749	**2 516**
1966	6 362	2 056	1 181	**9 599**
1967	11 891	4 989	2 513	**19 394**
1968	8 850	8 483	1 931	**19 264**
1969	12 376	3 572	1 309	**17 257**
1970	1 806	697	370	**2 873**
1971	–	38	. .	**38**
1972	–	–	–	**–**
1973	–	–	–	**–**
Total	**44 338**	**19 835**	**8 182**	**72 354**

Notes and sources:

(*a*) The agents are described in table 3.1.

(*b*) Data for 1962–67 are from Zablocki (1970:I:242). Data for 1968–69 are from Rivers (1970:8667:VII). Data for 1970 are from the US Library of Congress (1971:10). Data for 1971 are from Lang *et al.* (1974:S:3). The Orange/White partitions for 1966, 1967 and 1970 are in accordance with the proportions provided by Lang *et al.* (1974:S:3).

(*c*) Missing data have not been released by the US Department of Defense.

(*d*) To convert any of the above volume data to area covered in hectares, multiply by 35.6356. To convert any of the above Agent Orange volume data to 2,4,5-T content in kilograms, multiply by 545.448. Similarly for 2,4-D, multiply by 485.055, and for a dioxin estimate, multiply by 0.0025. To convert any of the above Agent White volume data to 2,4-D content in kilograms, multiply by 239.652. Similarly for picloram, multiply by 64.7060. To convert any of the above Agent Blue volume data to dimethylarsinic (cacodylic) acid in kilograms, multiply by 371.460. Similarly for elemental arsenic, multiply by 201.671. To convert any of the above total volume data to average kilograms of active ingredients, multiply by 756.918.

Table 3.3. US herbicide expenditures in the Second Indochina War: a breakdown by type of mission and agent [a]

$m^3 = 10^3$ litres

Type of mission[b]	Agent Orange	Agent White	Agent Blue	**Total**
Forest	39 816	19 094	1 684	**60 594**
Miscellanous woody vegetation	709	529	312	**1 550**
Crop	3 813	212	6 185	**10 210**
Total	**44 338**	**19 835**	**8 182**	**72 354**

Notes and sources:

[a] The several agent totals are from table 3.2.

[b] The partitions into type of mission are on the basis of the proportions provided by Lang *et al.* (1974:III:6). The "Forest" category here is equivalent to their "Defoliation" plus "Cache Site" categories. The "Miscellaneous Woody Vegetation" category here is equivalent to a combination of their "Perimeter", "Waterways", "Friendly Lines of Communication" and "Enemy Lines of Communication" categories. These data of Lang *et al.* take into account only fixed-wing aircraft herbicide missions and are based upon 92 per cent of total herbicide expenditures (96 per cent of Agent Orange, 100 per cent of Agent White, and 52 per cent of Agent Blue expenditures). In the table above, the quantity of Agent Blue unaccounted for in the data of Lang *et al.* (presumably that dispensed from helicopters) has all been assigned to the "Crop" category.

Orange was also much used for this purpose (table 3.3). All told, about 86 per cent of the missions were directed primarily against forest and other woody vegetation and the remaining 14 per cent primarily against crop plants. Total geographic coverage of the spray missions was less than one might expect on the basis of the total expenditure of herbicides since about 34 per cent of the target areas were chemically attacked more than once during the course of the war (table 3.4). Thus the total area subjected to spraying one or more times is estimated at 1.7×10^6 ha, this area being treated 1.5 times on the average, thereby receiving an average dose of about 42 litres/ha (or 32 kg/ha in terms of active herbicidal ingredients).

Most of the anti-plant chemicals (approximately 95 per cent) were dispensed from C-123 transport aircraft equipped to deliver somewhat over 3 600 litres onto 130 ha or so.[3] Most of the remainder was dispensed from helicopters, although small amounts were also dispensed via truck- and even boat-mounted spray rigs. The fixed-wing aircraft flew not far above tree-top level, at a relatively low speed, and had certain operational restrictions dealing with ambient wind and temperature conditions. These precautions were, however, insufficient to prevent at least some of the herbicides from landing off target. Among other things, the essentially unavoidable spray drift would seem to more than account for the fact that the area of herbicidal impact that has been reported by the recipient side was somewhat larger than that presented here.

Table 3.4. US herbicide expenditures in the Second Indochina War: a breakdown by number of repeat sprayings within the area covered

Number of sprayings of one area	Ultimate herbicide expenditure *m^3 = 10^3 litres*	Area involved *10^3 ha*
One	31 572	1 125
Two	21 431	382
Three	11 412	136
Four	5 335	48
Five or more	2 603	19
Total	**72 354**	**1 709**

Notes and sources:

(*a*) The total herbicide expenditure is from table 3.2.

(*b*) The partition among number of sprayings is based on data provided by Lang *et al.* (1974:III:24 & IV:54). The data of Lang *et al.* used for this computation take into account only fixed-wing aircraft herbicide missions. These data of Lang *et al.* are based upon 84.8 per cent of the total of all herbicide expenditures.

(*c*) The conversion of volume to area is on the basis of the standard rate of application of 28.062 litres/ha (table 3.1). Thereby ignored is the additional area subjected to herbicides by off-target drift of the chemicals.

(*d*) Had no area been sprayed more than once, then the total coverage would have been 2 578 x 10^3 ha. As it is, the areas which were sprayed received an overall average of 42.347 litres/ha, that is, they were sprayed an average of 1.509 times.

It is not possible to provide a very accurate regional breakdown of herbicide expenditures in Indochina inasmuch as the necessary information has not been released by the US Department of Defense (table 3.5). Whereas just over 2 per cent of all Indochina was sprayed one or more times this is a misleading statistic since most of the spraying was confined to South Viet-Nam, the total coverage of that country having been about 10 per cent. Within South Viet-Nam, it was Military Region III that was singled out for the most intensive coverage, most intensive both on a per unit area or *per capita* basis (map 1.2). Indeed, it appears that about 30 per cent of the land area of Military Region III was sprayed one or more times. This region was also the most heavily bombed and shelled (table 2.2) and the most widely cleared with tractors (*see* page 47).

III. *Ecological consequences*

The ecological consequences peculiar to large-scale herbicidal attack are the subject of both this subsection and the next.[4] The first step necessary is to approximate the extent and frequency with which various vegetational types

Table 3.5. US herbicide expenditures in the Second Indochina War: a breakdown by region

Region	Herbicide expenditure *m³ = 10³ litres*	Area sprayed once or more *10³ ha*	Proportion of area sprayed *per cent*	Volume sprayed per capita *litres/capita*
South Viet-Nam	72 354	1 709	9.9	4.1
Military Region I	12 300	290	10.3	4.0
Military Region II	15 194	359	4.7	4.9
Military Region III	38 348	906	30.0	7.9
Without Saigon				16.3
Military Region IV	6 512	154	4.0	1.0
North Viet-Nam	..	..	..	..
Cambodia	..	..	..	..
Laos	..	..	..	..
Total	**72 354**	**1 709**	**2.3**	**1.6**

Notes and sources:

(*a*) The total herbicide expenditure is from table 3.2. The total area sprayed once or more is from table 3.4.

(*b*) Amounts of herbicide expended in North Viet-Nam, Cambodia and Laos have not been separated out by the US Department of Defense from the total amount expended in Indochina, but are assumed to have been modest. The entire expenditure is, perforce, assigned here to South Viet-Nam.

(*c*) The partition among Military Regions is based on a breakdown of all fixed-wing herbicide sorties flown between mid-1965 and the end of 1970: M.R. I 17 per cent, M.R. II 21 per cent, M.R. III 53 per cent, and M.R. IV 9 per cent. The sample thus represents about 89 per cent of the total volume expended during the war (and represents roughly the final six years of spraying).

(*d*) The area and population data used are from table 1.1.

(*e*) To convert any of the above herbicide volume data to average kilograms of active ingredients, mulitply the cubic metres by 756.918, or the litres by 0.756 918.

were attacked. This is then followed by the effects of such attack on the autotrophic component or vegetation of the ecosystems involved, in the present subsection particularly of upland (inland) ecosystems. Discussed next is the effect on the nutrient budget of herbicidally attacked ecosystems, followed in turn by considerations of the heterotrophic components, both the larger animals and microorganisms. The present subsection is concluded with an examination of the environmental persistence of the herbicides which were used in Indochina, and thus of the extent of their residual effects. The special case of herbicidal attack on the coastal mangrove habitat is reserved for the next subsection (*see* page 38). The more general and long-range aspects of herbicidal disruption of an ecosystem are discussed elsewhere (chapter 6).

It can be estimated that about 14 per cent of the total extent of South Viet-Nam's woody vegetation has been sprayed one or more times (table 3.6). With respect to the dense upland forest type in its various stages of succession (comprising perhaps 57 per cent of the total), the extent of attack has been estimated to be somewhat greater, about 19 per cent.

Table 3.6. US herbicide expenditures in the Second Indochina War: a breakdown by vegetational type

Vegetational type (South Viet-Nam)	Area of type	Area sprayed once or more
Dense forest	5 800	1 110
Pure dense forest	4 500	861
Pure plus secondary dense forest	600	115
Secondary dense forest	700	134
Open (clear) forest	2 000	100
Bamboo brake	800	40
Mangrove forest (swamp)	500	151
True mangrove	300	124
Rear mangrove	200	27
Rubber plantation	100	30
Pine forest	100	–
Miscellaneous woody (brush, etc.)	1 100	37
Woody subtotal	**10 400**	**1 467**
Paddy (wet) rice	2 500	60
Field crops (upland rice, etc.)	500	181
Agricultural subtotal	**3 000**	**241**
Miscellaneous	3 926	–
Total	**17 326**	**1 709**

Notes and sources:

(*a*) Areas of the vegetational types are from table 1.2.

(*b*) The total area sprayed once or more is from table 3.4. It is somewhat in error as used here for South Viet-Nam inasmuch as the value represents expenditures directed against all of Indochina. However, the modest but unknown amounts sprayed onto Cambodia, Laos and perhaps North Viet-Nam have not been released by the US Department of Defense.

(*c*) The total area sprayed once or more is fractionated as follows: The first partition is into the three categories provided in table 3.3: forest 83.7 per cent, miscellaneous woody vegetation 2.1 per cent, and agricultural (crop) vegetation 14.1 per cent. The areas sprayed in the two mangrove sub-types are based on those of Lang *et al.* (1974:III:24), increased to account for the 84.8 per cent sample upon which these authors based those figures. The areas sprayed in the following types are the more or less arbitrary fractions of the total area of each type given: open forest 5 per cent, bamboo brake 5 per cent, rubber plantation 30 per cent, and pine forest 0 per cent. The remaining as yet unassigned area of sprayed woody vegetation is assigned to the dense forest type, being split among its sub-types in proportion to their areas. The agricultural (crop) spraying is partitioned arbitrarily 25 per cent paddy and 75 per cent field.

(*d*) Rubber = *Hevea brasiliensis,* Euphorbiaceae. Pine = *Pinus,* Pinaceae. Rice = *Oryza sativa,* Gramineae.

The 1.1 x 10^6 ha of dense forest that were sprayed in War Zones C and D and elsewhere in South Viet-Nam are composed of an incredibly complex floristic mixture. When such a dense upland forest is subjected to herbicidal attack, this results in fairly complete leaf abscission (as well as flower and fruit abscission) within two or three weeks (Darrow *et al.*, 1971). The surviving trees usually remain bare until the onset of the next rainy season, often therefore, for a period of several months. To achieve the total defoliation of the lower storeys or to prolong the period of leaflessness may necessitate a follow-up spraying.

Virtually all of the endless number of dicotyledonous tree species are defoliated at the intensity of spraying employed. Then at the time of refoliation it becomes evident that there is a spectrum of sensitivity among the many hundreds of tree species which comprise the floristically complex and variable dense upland forest type. Only about 10 per cent of the trees are killed outright by a single military spraying, a situation presumably obtaining on 66 per cent of the total sprayed area (table 3.4). The remainder display various levels of injury, as evidenced by differing severities of crown (branch) dieback, temporary sterility, and other symptoms. Among the most sensitive of the dense forest species of South Viet-Nam are *Pterocarpus pedatus* (Leguminosae) and *Lagerstroemia* spp. (Lythraceae). Among the most resistant are *Cassia siamea* (Leguminosae) and *Sandoricum indicum* (Meliaceae). And among those intermediate between these two extremes are *Hopea odorata, Dipterocarpus alatus,* and *Shorea cochinchinensis* (all Dipterocarpaceae).

When the military situation leads to more than one herbicidal attack (as occurred on about 34 per cent of all the sprayed lands), the level of tree mortality increases with each subsequent spraying (more so with briefer intervals between sprayings). Two herbicidal attacks, as occurred on about 22 per cent of the sprayed lands, result in an estimated average mortality rate of 25 per cent. Three such attacks, as occurred on about 8 per cent of the sprayed lands, result in an estimated average mortality rate of 50 per cent. And four or more such attacks, as occurred on the remaining 4 per cent of the sprayed lands, result in estimated mortality rates ranging from 85 per cent to essentially 100 per cent.

Unlike a temperate-forest ecosystem, or even an arid tropical one, the major fraction of the total nutrient budget of tropical upland forest ecosystems of the sort found in South Viet-Nam is found at any one time in their biotic component, much of it in the leaves and small twigs (*see* page 68). Thus a very small fraction is in the soil component of these systems. Leaves and other plant and animal remains drop to the ground more or less steadily throughout the year. This litter decomposes rapidly, thereby releasing its nutrients for recycling into the living vegetation. Because of the rapid decomposition of the organic matter and the high level of rainfall, the soil here does not have a high capacity to retain nutrients, but continues to supply them to the vegetation from current acquisitions. A very efficient (closed) biogeochemical cycle is in operation, albeit a fragile one.

It is thus safe to suggest that when chemical agents cause a major leaf fall in an Indochinese upland forest ecosystem, this nutrient-rich litter will decompose rapidly, the vegetation will be dormant or moribund, and the released nutrients will be largely lost to the area via rapid leaching, runoff and erosion. This phenomenon, termed "nutrient dumping", has, in fact, been demonstrated to occur following herbicidal attack in South Viet-Nam, and is discussed more fully on page 66.

Herbicidal attack has a serious impact not only on the autotrophic component of an ecosystem described earlier (that is, the first link in any food chain), but on the higher level, heterotrophic links as well. Simply put, the animals – large and small – expect food and cover from their *milieu*. A significant majority of the animal life in a tropical forest is found in the upper vegetational storeys, precisely the portion of the ecosystem most seriously impaired by massive herbicidal attack. With a destroyed or at least drastically altered habitat, one is led to expect a concomitant change of catastrophic proportions in the animal populations. For example, Hull (1971) was able to observe a major reduction in population numbers of the gopher *Thomomys talpoides* (Geomyidae) following treatment of an area in Idaho, USA, with 2,4-D. Hull attributed this decimation entirely to the herbicidal destruction of the food supply of this herbivore (*see also* Keith *et al.*, 1959; Tietjen *et al.*, 1967). Similarly, Zimin (1971) found that nesting birds vacated a conifer forest in northwestern USSR in which the dicotyledonous component had been killed with 2,4-D, and that mammalian numbers also declined in the area. Pate *et al.*, (1972) and Young (1974:IV) were able to attribute changes in various vertebrate populations on a herbicide test site in Florida, USA to changes in the vegetation. Furthermore, Valder (1972) and Young (1974:V) correlated the disappearance of insects from this same site with the disappearance of plant species used by them for food or shelter (*see also* Oliver *et al.*, 1966).

Beyond their debilitating effect on wildlife via a destruction of their food and cover, herbicidal agents can also be directly toxic to exposed animals.[5] All of the agents that were used in Indochina are known to be toxic to animal life at sufficiently high dose rates. Thus, Zimin (1971) in northwestern USSR has observed impaired reproductive abilities in forest birds subsequent to a silvicultural treatment with 2,4-D of a conifer forest (*see also* DaLage & Alnot, 1973; Lutz-Ostertag & Lutz, 1970). Lofroth (1970) reported large numbers of deaths and abortions among reindeer (*Rangifera tarandus*, Cervidae) in northern Sweden that were forced by weather conditions to feed on brush which had been aerially treated some months previously with a 2:1 mixture of 2,4-D and 2,4,5-T, sprayed at the rate of 2 kg/ha (*but see* Erne, 1973). Swiggart *et al.*, (1972) reported the deaths of numerous deer (*Odocoileus virginianus*, Cervidae) following aerial field spraying in Tennessee, USA of about 240 ha with arsenic acid at about 64 per cent of the military dose rate for Agent Blue, in terms of their respective arsenic contents (table 3.1). Arsenic acid is an inorganic herbicide rather similar in structure to the dimethyl arsinic acid comprising Agent

Blue (its arsenic also being in the pentavalent state, but possessing hydroxyl groups *in lieu* of the two methyl groups). It is also of interest to note here that Sjödén & Söderberg (1972; 1975) have observed behavioural abnormalities in the offspring of rats (*Rattus rattus,* Muridae) treated during pregnancy with sublethal dose levels of 2,4,5-T.

Although there is little direct evidence from Indochina of wildlife decimation that can be attributed to herbicidal attack, the above-cited instances should make it clear that it was likely to have occurred. Local inhabitants in Cambodia have described how some wild birds of the forest became disabled for a time following herbicidal attack so that they could be captured easily, and that some of the smaller ones died (Westing, 1972a:196; *see also* C. Carter *et al.,* 1975). Moreover, the use of herbicides may have placed in jeopardy the actual survival of two animal species in South Viet-Nam: the freshwater tarpon *Megalops cyprinoides* (Megalopidae) and the estuarine clam *Polymesoda coaxans* (Corbiculiidae) (*see* page 72).

Turning now to invertebrates, a number of kinds of such organisms might have been poisoned by the herbicides used in Indochina. For example, various aquatic invertebrates are known to be sensitive to the herbicides employed (Butler, 1965; Sanders, 1970). The loss of these as a source of food may have contributed to the observed declines in numbers of fish (*see* page 34). Although the work of C. Fox (1964) in Nova Scotia, Canada suggests that 2,4-D has little effect on certain soil inhabiting invertebrates, various of the terrestrial insects in Indochina are likely to have been poisoned in substantial numbers, including some of probable importance as pollinators. It is known from civil experience that the honey bee (*Apis mellifera,* Apidae) is moderately sensitive to 2,4-D and 2,4,5-T and extremely sensitive to dimethyl arsinic acid (Moffett *et al.,* 1972; Morton *et al.,* 1972; Palmer-Jones, 1964), and thus other Hymenoptera could be affected in like manner. Vinegar flies (*Drosophila melanogaster,* Drosophilidae), and therefore probably other Diptera as well, are susceptible to injury by both 2,4-D and 2,4,5-T (Dävring & Sunner, 1971; Novy & Majumdar, 1972). Moreover, various Scolitidae, and for that reason perhaps other Coleoptera, readily succumb to low doses of dimethyl arsinic acid, among them species of *Dendroctonus* (Buffam *et al.,* 1973; Chansler *et al.,* 1970; Frye & Wygant, 1971), *Ips* (Frye & Wygant, 1971; Stelzer, 1970), *Polygraphus, Scierus* (Frye & Wygant, 1971), *Scolytus,* and *Pseudopityophthorus* (Rexrode & Lockyer, 1974). Finally, the Coleoptera *Coccinella* and *Hippodamia* (both Coccinellidae) are highly susceptible to 2,4-D poisoning in their larval stages (J. Adams, 1960; J. Adams & Drew, 1965; 1969). In fact, when this herbicide was used in New Brunswick, Canada to control weeds in fields of oats (*Avena sativa,* Gramineae) or barley (*Hordeum vulgare,* Gramineae), aphid (Aphididae) populations normally kept in check by the coccinellids rose dramatically (and in turn had to be controlled by the application of an insecticide).

The next question that arises is the level to which the herbicidal attacks in Indochina affected the diverse microorganisms of the sprayed ecosystems. Of

particular potential concern would be the effects on those soil microorganisms involved in the nitrogen cycle, in the sulphur cycle, in mycorrhizal relationships and in the decomposition of litter.

In fact, it appears from the literature that the effects on the involved microorganisms were at worst modest or transitory.[6] This is particularly so in the light of the reproductive capacities of these forms of life and of the relatively brief environmental persistence of the herbicides used (*see* below). It will be useful to cite some of the studies upon which this conclusion is based.

With respect to phenoxy herbicides, Breazeale & Camper (1970) found that soil applications of 2,4-D greatly in excess of the military ones[6] had no discernible effect on total population counts of either fungi or actinomycetes, but seem to have depressed the total population level of the bacteria other than the actinomycetes (*see also* Young, 1974:VII). On the other hand, Arnold *et al.* (1966) found that 2,4-D applications in excess of military ones inhibited the proliferation of the ascomycete *Aspergillus niger* (Aspergillaceae). Moreover, Liu & Cibes-Viade (1972) found that 2,4-D in excess of military dose levels somewhat reduced the total activity of microorganisms of a tropical soil, as measured by an inhibition of oxygen uptake (the so-called soil respiration). Foldesy *et al.* (1972) observed adverse effects on two nitrogen-fixing bacteria (*Azotobacter*, Azotobacteriaceae and *Rhizobium*, Rhizobiaceae) from 2,4,5-T concentrations applied in great excess of military application levels. Fletcher & Raymond (1956) have reported a suppression of *Rhizobium* nodulation in the roots of *Trifolium* (Leguminosae) as a result of 2,4-D and particularly of 2,4,5-T applications to the soil at approximately military dosages. Warren *et al.* (1951) found that although 2,4-D treatment caused an overall reduction in the soil microflora, there was a conspicuous increase in one particular species of actinomycete.

With respect to picloram, Dubey (1969) found that a dose level in excess of the military ones caused a mild inhibition of the nitrification capacity of a tropical soil (that is, its ability to oxidize the NH_4^+ ion to NO_2^- and NO_3^- ions) (*see also* Grover, 1972). Arnold *et al.* (1966) found that picloram in excess of military application levels had no effect on the proliferation of *Aspergillus niger* (Aspergillaceae). Regarding dimethyl arsinic acid, Zabel & O'Neil (1957) could not inhibit the growth of two bacteria (*Bacillus mycoides*, Bacillaceae and *Aerobacter aerogenes*, Enterobacteriaceae) and two fungi (*Penicillium expansum* and *Aspergillus niger*, both Aspergillaceae) using dosages far in excess of military ones. Bollen *et al.* (1974) found that dimethyl arsinic acid applications to forest soil in excess of the military dosages had no pronounced effect on microbial organic matter decomposition (as measured by the evolution of CO_2).

Turning now to another matter, a small but significant fraction of herbicides applied to terrestrial targets is unavoidably sprayed onto or washed into the runoff water (Trichell *et al.*, 1968). Once in the runoff, the herbicides find their way into the streams, the estuaries and the ocean. Particularly the hormonal herbicides that were used so lavishly in Indochina are known to be lethal to many aquatic plants and at least moderately toxic to numerous species of true fish ("finfish") and crustaceans ("shellfish"), as well as to various of the lower organisms used by these two groups for food.[7] It thus becomes evident that not only are the terrestrial ecosystems disrupted in a herbicidally attacked region, but also the aquatic ones. The involved aquatic fauna is debilitated both directly (via poisoning) and indirectly (via food and habitat disruption). For example, the widespread disruption of the South Viet-Namese mangrove estuaries seems to have had a major impact on the numerous fish and crustaceans that depend upon this habitat for their breeding or nursing grounds during certain periods of the year or perhaps even the year round (*see* below).

Important in any consideration of the ecological impact of herbicides is their persistence and mobility, that is, how long they will remain present and active in the soil and biota and whether they will move up in the food chain, perhaps even becoming concentrated in the process.[8] During an extended tenure in the ecosystem, herbicide residues could adversely affect the replacement vegetation, soil microorganisms and other flora and fauna.

The rate of disappearance of any particular herbicide depends upon a complexity of factors. Important among the intrinsic ones are the chemical's vulnerability to microorganismal decomposition as well as its volatility, its photo- and thermo-degradabilities, and its adsorptive (binding) properties and related leachability. Important among the extrinsic factors are the ambient temperature regime, abundance of rainfall, soil type and recent spray history of the site. Initial dose rate is not quite as important in this regard as one might expect since for many herbicides the observed rate of disappearance turns out, after a brief initial lag period, to be proportional at any one time to the herbicidal concentration in the soil (thus describing a so-called first-order reaction).

At civil doses of application in temperate regions, dimethyl arsinic acid is usually considered to reach a level of insignificance within a week or so, as determined by lack of obvious effect on subsequently planted crops. For 2,4-D this level is reached within a month or so, for 2,4,5-T within five months or so, and for picloram within 18 months or so.[8] Chemical analysis would, of course, reveal traces of all these substances for some time beyond the durations just given, as might the sowing of particularly sensitive indicator species. Some specific examples from the literature follow, taken so far as is possible from experiences in the tropics (*see* table 3.1 for military dose rates).

In Hawaiian field trials, Akamine (1950–1951) found that 2,4-D applied at 11 kg/ha continues to affect subsequently planted indicator plants such as bean (*Phaseolus,* Leguminosae) or tomato (*Lycopersicon,* Solanaceae) for two to 14 weeks, depending upon soil and other environmental factors. In some military field trials in Maryland, USA with herbicidal concentrations up to 30 kg/ha, Newman *et al.* (1952) found that 2,4-D levels became similarly insignificant within six weeks, but that for 2,4,5-T it took 19 weeks. Bovey *et al.* (1968) in Puerto Rican field trials found that a 1:1 mixture of 2,4-D and 2,4,5-T applied at 27 kg/ha exerted a significant residual effect on various herbaceous monocotyledons and dicotyledons for two months, and picloram at 7 kg/ha for three months. They attributed the speed of soil rehabilitation to high local rainfall. In further trials there, Bovey *et al.* (1969b) could detect the effects of picloram applied at 3 kg/ha on cucumber (*Cucumis,* Cucurbitaceae) for 12 months, results also comparable to those of Dowler *et al.* (1968). (*See also* Meikle *et al.,* 1973). By way of contrast, VandenBorn (1969) found that in the cool, low-rainfall climate of Alberta, Canada, the effects of picloram applied at 3 kg/ha are still evident on sensitive species such as sunflower (*Helianthus,* Compositae) planted five years subsequent to the herbicidal application. Moreover, in northern Sweden 2,4-D and 2,4,5-T residues could be detected in the tissues of a variety of higher and lower plants, game animals and fish for up to two years after a single forest spraying at 5 kg/ha or less (Erne, 1974; Erne & Haartman, 1973).

The poisonous dioxin (2,3,7,8-tetrachlorodibenzo-*p*-dioxin) contaminant found in Agent Orange[9] has been found to be highly persistent in soil (Helling *et al.,* 1973). When Kearney *et al.* (1972) in laboratory experiments applied dioxin to two diverse soils at dose rates approximately equivalent to 2.65 kg/ha or higher, recovery of the dioxin after one year was 56 per cent for one of the soils and 63 per cent for the other. However, Isensee & Jones (1971) concluded on the basis of their studies that the accumulation of dioxin by plants from soil was highly unlikely. The ecological significance of dioxin added to a terrestrial ecosystem is still obscure (*see also* Kearney *et al.,* 1973a; Kearney *et al.,* 1972; Kearney *et al.,* 1973b; Norris & Miller, 1974).

Applied dimethyl arsinic acid appears to become for the most part rapidly unavailable to subsequently planted vegetation via adsorption (binding) onto soil particles. For example, Ehman (1965) has reported that for such horticultural operations as lawn renovation, grass (Gramineae) seed can be sown without fear of harm to it immediately following an application of 22 kg/ha, since sufficient inactivation occurs during the few days prior to seed germination (*see also* Johnson & Hiltbold, 1969; Misra & Tiwari, 1963). Ehman (1965) further reported that subsequently harvested alfalfa (*Medicago,* Leguminosae) and rye-grass (*Lolium,* Gramineae) that had been sown three days after pasture treatment with 6 kg/ha contained arsenic at levels no higher than in control plants.

Through time, dimethyl arsinic acid disappears from soil to a large extent via its microorganismal conversion to the gas arsine (AsH_3) or related compound. Such conversion appears to be common in the forest environment (S. Wagner & Weswig, 1974) and is known to be mediated both by fungi (Challenger, 1945; Cox & Alexander, 1973; Thom & Raper, 1932) and bacteria (McBride & Wolfe, 1971). Woolson & Kearney (1973) have reported soil losses of this sort that ranged between 35 and 61 per cent during a period of 24 weeks.

It is important to note in the present context that all soils contain a natural level of arsenic, some of it even in the form of dimethyl arsinic acid (Braman & Foreback, 1973). The most frequent natural soil levels of arsenic encountered fall between 5 mg/kg and 10 mg/kg, but can be considerably higher (K. Williams & Whetstone, 1940). Military applications supply of the order of 2 mg/kg.[6] There is a natural arsenic cycle in nature (and arsenic may even be an essential trace element) (Frost, 1967; J. Wood, 1974). There seems to be no food-chain concentration (so-called ecological magnification) of arsenicals to toxic levels (Isensee *et al.*, 1973; Schuth *et al.*, 1974).

With respect to investigations of herbicidal persistence in South Viet-Nam itself, Minarik & Darrow (1968) reported that there was no discernible effect on beans (*Phaseolus*, Leguminosae) germinated and grown for nine days in soils that had been subjected to two military sprayings either with Agent Orange (collected 16 months after the sprayings) or Agent White (collected 11 months after the more recent spraying) (*see also* Blackman *et al.*, 1974:17–19). Moreover, in field trials in South Viet-Nam, Blackman *et al.* (1974) found that when Agent Orange was sprayed onto cleared upland forest soils at military dose levels, such treatment ceased to cause a reduction in survival and growth of, *inter alia*, subsequently planted corn (*Zea*, Gramineae) after four weeks, of upland rice (*Oryza*, Gramineae) or peanut (*Arachis*, Leguminosae) after 10 weeks, and of bean (*Phaseolus*, Leguminosae) after 18 weeks. In an equivalent trial with Agent White the times to insignificant damage were 10 weeks for both corn and upland rice, 31 weeks for the bean used, and 24 weeks for peanut (although some presumably trivial effects were still discernible on the peanuts for at least 34 weeks, the time at which the observations were terminated). When either Agent Orange or White was similarly applied to cleared mangrove forest soils, Blackman *et al.* (1974) reported that the herbicidal rate of disappearance was similar to that found in the upland forest soils. Moreover, when seedlings of two common mangrove species (*Rhizophora* and *Ceriops*, both Rhizophoraceae) were transplanted into such treated soils 40 days after spraying, their rate of survival appeared to be equivalent to that in control soils.

Thus, from all of the above one can conclude that under the hot and rainy conditions of Indochina the military herbicides selected for use probably did not exert a substantial ecological influence via residual persistence.

A major consideration in the ecological impact of herbicidal attack is the long-term changes in the biotic community brought about by the decimation of the vegetation. Particularly in those areas in South Viet-Nam that were sprayed several times – some 200 x 10^3 ha (table 3.4) – the overstorey destruction was sufficient to permit the release of or invasion by certain tenacious pioneer grasses (Gramineae). These include both herbaceous types, such as the notorious *Imperata cylindrica*, and woody types such as *Bambusa*, *Thyrsostachys*, *Oxytenanthera*, and other frutescent (shrubby) bamboos. The implications of these occurrences are discussed elsewhere (*see* page 78).

Thus it can be seen that the ecological impact of military spraying, particularly if repeated, is a severe one on upland forests. The primary producers are knocked back drastically, to be replaced by a new plant community of significantly lesser biomass and smaller nutrient-holding capacity. A poorer soil results from the attack, one with a lesser fraction of humus (organic matter) and one that also often exhibits a chronic shortage of nitrogen (*see* pages 65 and 67). Fire following defoliation would aggravate the entire situation (*see* page 67). Moreover, the primary productivity of the forest is reduced by these events, and with it the entire energy base for the ecosystem. There exist a number of more or less subtle additional ways in which such a herbicidally attacked ecosystem might be debilitated. These include the decimation of certain plants particularly important as symbionts for nitrogen-fixing microorganisms (Youngberg, 1965), the sparing of other plants that might favour the spread of disruptive virus diseases (Milinkó, 1967) and the metabolic disturbance of yet other plants, causing them to become toxic to the herbivores feeding on them (Stahler & Whitehead, 1950; C.R. Swanson & Shaw, 1954; E. Whitehead *et al.*, 1956; Willard, 1950).

The mangrove habitat, scattered primarily along the southern coastline of South Viet-Nam, occupies approximately 0.5 x 10^6 ha of inhospitable, seemingly impenetrable, and outwardly unimportant swamps. It is singled out here owing to the widespread and peculiarly drastic damage it suffered during the war from herbicidal attack.[10]

An estimated 124 x 10^3 ha of true mangrove (41 per cent of that entire subtype) plus another 27 x 10^3 ha of rear mangrove (13 per cent of that subtype) were subjected to military herbicide spraying during the Second Indochina War (table 3.6). Unlike the relatively modest degree of kill resulting from such action in the upland forest types, even a single spraying in this coastal lowland type most often destroys essentially the entire plant community. Virtually nothing remains alive after a single herbicidal attack and the resulting scene is a strange and desolate one. It subsequently becomes even more so when exacerbated by the usual salvage harvesting of the killed trees and by the inevitable erosion.

The taxonomically diverse plant species that make up the mangrove community all display great sensitivity to the hormone-mimicking herbicides, with *Rhizophora* (Rhizophoraceae) being especially sensitive (Minarik & Bertram, 1962; Truman, 1961–1962; Walsh *et al.*, 1973). It almost seems as if the physiological attributes which permit these species to thrive under the constraints of

---

Facing page:

Ecological effects of high-explosive munitions, I: B-52 strike against a lowland forest area in Bien Hoa province, South Viet-Nam, photographed on 8 August 1971

"With a strike zone of 65 ha per B-52 aircraft and an average of 4.6 aircraft per Indochina mission, the average zone of devastation per B-52 mission can be seen to have been 296 ha . . . The total area carpeted by B-52s . . . was 8.1 x 10^6 ha – just over 11 per cent of the total area of Indochina. If one singles out South Viet-Nam (a region accounting for 53 per cent of the B-52 bomb tonnage expenditures), the comparable data . . . add up . . . to fully 26 per cent of South Viet-Nam's area." (p. 16)

Ecological effects of high-explosive munitions, II: Aerial view of craters in a lowland forest area in Bien Hoa province, South Viet-Nam, 8 August 1971

"A huge amount of soil . . . was displaced by the high-explosive munitions expended in Indochina. A rough calculation suggests this amount to have been of the order of 3×10^9 m^3 . . . Under ordinary field conditions the longevity of a crater appears to be many years or decades . . . It becomes possible to calculate that roughly 21×10^6 bombs and 229×10^6 shells were expended [in Indochina] – and thus, that there now exist almost the same number of permanent craters, large and small." (p. 21)

Ecological effects of anti-plant chemicals, I: Aerial view of forest area in Long Khanh province, South Viet-Nam, 8 August 1971

"Forest destruction was generally accomplished through the use of Agents Orange or White, whereas Agent Blue was usually the agent of choice for the destruction of rice . . . and other crops . . . All told, about 86 per cent of the missions were directed primarily against forest and other woody vegetation and the remaining 14 per cent primarily against crop plants . . . The total area subjected to spraying one or more times is estimated at 1.7×10^6 ha, this area being treated 1.5 times on the average . . . Most of the spraying was confined to South Viet-Nam, the total coverage of that country having been about 10 per cent." (pp. 26–28)

Ecological effects of anti-plant chemicals, II: Aerial view of forest area in War Zone D, Long Khanh province, South Viet-Nam, 8 August 1971

"There is a spectrum of sensitivity among the many hundreds of tree species. . . Only about 10 per cent of the trees are killed outright by a single military spraying, a situation. . . obtaining on 66 per cent of the total sprayed area. . . Two herbicidal attacks, as occurred on about 22 per cent of the sprayed lands, result in an estimated mortality rate of 25 per cent. Three such attacks, as occurred on about 8 per cent of the sprayed lands, result in an estimated average mortality rate of 50 per cent. And four or more such attacks, as occurred on the remaining 4 per cent of the sprayed lands, result in estimated mortality rates ranging from 85 per cent to essentially 100 per cent." (p. 31)

Ecological effects of anti-plant chemicals, III: Mangroves in Gia Dinh province, South Viet-Nam, 15 August 1970, destroyed by spraying several years before

"An estimated 124×10^3 ha of true mangrove (41 per cent of that entire subtype) plus another 27×10^3 ha of rear mangrove (13 per cent of that subtype) were subjected to military herbicide spraying. . . Virtually nothing remains alive after a single herbicidal attack. . . Little if any recolonization occurs on herbicide-obliterated mangrove sites. . . Some 30 per cent of South Viet-Nam's once lush mangroves [have been converted] to a muddy wasteland for an indefinite period of time." (pp. 38–39)

Ecological effects of mechanized landclearing, I: Aerial view of land cleared by Rome ploughs in the Boi Loi woods, Tay Ninh province, South Viet-Nam, 10 August 1971

"Some 325×10^3 ha were cleared by the Rome ploughs in South Viet-Nam, that is, 2 per cent of its entire land area. . . Obliterated there, for example, were the 7 100-ha Hô Bô woods in central Bien Hoa province, the 3 600-ha woods of the same name in west-central Binh Duong province and the 2 700-ha Boi Loi woods in southeastern Tay Ninh province." (p. 47)

Ecological effects of mechanized landclearing, II: Aerial view showing craters and area cleared by Rome ploughs in Tay Ninh province, South Viet-Nam, 10 August 1971

"The damage to the site from erosion can be massive during the months before a new vegetational cover becomes established... The landclearing operations also bring about additional, more subtle soil degradation via the loss in solution of mineral nutrients... There is an aggravation of flooding problems... The exceedingly rich tropical faunal community that depends for its food and shelter on the forest cover... is, of course, decimated... Mechanized landclearing operations lead to a remarkable alteration of the microclimate... Many years can be expected to elapse before natural forces will restore a Rome-ploughed site to its former status." (pp. 48–49)

War damage to South Viet-Nam's forests: Aerial view showing herbicidal and bomb destruction, Gia Dinh province, South Viet-Nam, 14 August 1970

"Combining the several separate damage estimates. . . for the forests of South Viet-Nam is made difficult by an inability to determine with any degree of accuracy the extent to which any one area was subjected to more than one of the several categories of insult. . . Complete or essentially complete devastation comes to an estimated 568×10^3 ha, representing 5 per cent of the total forest lands. . . War damage examined here amounted to a merchantable timber loss of about 74×10^6 m^3, that is, to about 14 per cent of the nation's standing merchantable timber crop." (pp. 70–71) the tropical tidal zone are in some way linked to their herbicidal sensitivity. The only partial exception to this widespread sensitivity in the mangrove community is *Avicennia* (Verbenaceae) growing on the edge of small watercourses. Individual trees in this category sometimes survive a spraying.

Again for reasons that seem elusive, little if any recolonization occurs on herbicide-obliterated mangrove sites. However, it appears that a lack of adequate seed source, a destruction of available propagules by crabs (Brachyura), and other more obscure factors have combined to cause this, that is, to convert an important fraction of South Viet-Nam's once lush mangroves – some 30 per cent of them – to a muddy wasteland for an indefinite period of time. Indeed, it has recently been estimated that without a concerted replacement and protection effort by man, substantial recovery of South Viet-Nam's spray-devastated mangroves can be expected to take more than a century (Lang *et al.*, 1974:IV:119; H. Odum *et al.*, 1974:289).

The lack of mangrove recolonization appears not to stem from any persistence of herbicide residues in the soil. As mentioned previously, Blackman *et al.* (1974) transplanted seedlings of *Rhizophora* and *Ceriops* (both Rhizophoraceae) into soils 40 days after they had been subjected to herbicidal treatment at military dose levels. Their rate of survival appeared to be equivalent to that in unsprayed control sites. That the sheer magnitude of the devastation may be in part responsible for the continued desolation is suggested by the natural recolonization by *Rhizophora* and *Avicennia* (Verbenaceae) which has been observed near the coast of the Gia Dinh province in a small clearing associated with an abortive attempt at mariculture.

The ecological impact of essentially total and long-term vegetational destruction in the mangrove type is rather obvious in some respects and more subtle in others. With the primary producers essentially wiped out, their energy-capturing function is lost to the ecosystem – and soon all else that builds on this. Both food and cover were eliminated by the US attacks not only for its enemy forces, but for most other indigenous creatures as well. For example, Orians & Pfeiffer (1970) reported an enormous reduction in the number of birds (*see also* Cook, 1970–1971; Westing, 1971b).

Less obviously, the endless reticulation of channels throughout a mangrove swamp (roughly one-fourth of the surface area of the mangrove type in South Viet-Nam) supports a rich variety of aquatic fauna during all or part of their life cycle. These organisms depend directly or indirectly on a steady and enormous supply of nutrients dropped, flushed and leached out of the terrestrial part of the system (Heald & Odum, 1970). Numerous species of fish and crustaceans that spend their adult lives offshore, and some that migrate up the rivers, utilize the mangrove estuaries as breeding and/or nursery grounds. Indeed, it is now known that various aquatic animal populations have been depleted as a result of the mangrove destruction. For example, various snails and other molluscan species have at least temporarily vacated the herbicide-destroyed mangrove areas, presumably owing to the remarkably altered local site conditions (G. Davis, 1974; Lang *et al.*, 1974:IV:117). G. Davis's (1974) findings, furthermore, suggest that one species of mollusc might have been placed in direct jeopardy by the herbicides. He found population levels of the clam *Polymesoda coaxans* (Corbiculiidae) to be dangerously depleted. Moreover, the few remaining survivors were apparently no longer reproducing, suggesting direct agent toxicity. Although the herbicides in question are not noted for their molluscan toxicity (Pimental, 1971), picloram may be an exception (Lynn, 1964–1965). As to other forms of life in the herbicide-destroyed mangrove areas, Sylva & Michel (1974) and Lang *et al.* (1974:IV:113) were able to note significant decreases in the number and variety of planktonic and benthic forms (diatoms, copepods, and so on) as well as in fish eggs. Declines have also been noted in South Viet-Nam's offshore fishery, involving both true fish and shellfish (crustaceans) (Brouillard, 1970:21; Loftas, 1970; Sylva & Michel, 1974:113–17).

Finally it must be mentioned that the mangrove type is a transitional zone between land and sea and thus appears to serve the added important function of stabilizing the shoreline. As the coastline accretes, mangroves invade the virgin lands and their roots hold the soil against the actions of wind, wave, current and tide. There are strong indications that both the now unprotected channel banks (Sylva & Michel, 1974:121) and barren mud flats created by the spraying are eroding away at rapid rates (Ross, 1974:30–31; Weatherspoon & Krusinger, 1974b:23–24). Indeed, Weatherspoon & Krusinger have found evidence for accelerated sheet erosion which has resulted in widespread lowering of the land surface of the now barren mangrove sites. More than 10 cm was reported to have been lost in this way merely during the first several years after spraying.

Thus it is evident that the spraying of the mangroves of Indochina has had a drastic impact on this semi-aquatic, tropical estuarine ecosystem – an ecosystem considered to be among the most highly productive in the world (Golley *et al.*, 1962) – and that the ramifications of this assault will be long and widely felt. These ramifications will include not only the various ecological and geomorphological ones alluded to above, but also economic problems including forestry and fishery losses and even exacerbated public health problems.

IV. *Conclusion*

Chemical anti-plant warfare, although not an innovation of the Second Indochina War, was during this conflict employed at such a profligate level that its use has become inseparably associated with it. Moreover, military evaluations have been sufficiently favourable so that it would come as no surprise if forests or crops were to be destroyed in a similar fashion during some future conflict (*see* page 84).

Among the ecological lessons to be learned from the chemical anti-plant warfare in Indochina are: (*a*) that the vegetation can be utterly destroyed with relative ease over extensive areas; (*b*) that this in turn has a devastating impact on the animal life depending upon this vegetation for food or shelter; (*c*) that the ecosystem is, through such attack, subject to nutrient dumping (that is, to the rapid and major loss of soluble nutrients); and (*d*) that the ecological debilitation from such attack is likely to be of long duration. To this list can be added a number of social lessons to be learned as well, among them: (*a*) that natural, agricultural and industrial-crop plant communities are all similarly vulnerable, and (*b*) that the local civil population can suffer extensively from such action in a variety of direct and indirect ways.

Notes

[1] For an introduction to the voluminous literature on the mode of action of chemical anti-plant agents or herbicides, *see* Ashton & Crafts (1973) and Corbett (1974). For methods of application, *see* note 2.

Several hundred publications dealing with the military use of chemical anti-plant agents exist (Westing, 1974a). McConnell (1969–1970), a representative of the US Department of Defense, has prepared a descriptive article on the subject (*see also* Westing, 1971–1972b). Moreover, the US Department of Defense has commissioned two open studies that deal with the military use of herbicides, those by House *et al.* (1967) and by the US National Academy of Sciences (1974). The latter was a major study with a report more than 1 800 pages in length (*see,* for example, Lang *et al.*, 1974; Blackman *et al.*, 1974; G. Davis, 1974; Desowitz *et al.*, 1974; Drew, 1974a; 1974b; Fryer, 1974; H. Odum *et al.*, 1974; Ross, 1974; Sylva & Michel, 1974; Weatherspoon & Krusinger, 1974a; 1974b; Zinke, 1974). Finally, considerable information on the use and implications of herbicides as chemical warfare agents has been incorporated by the Stockholm International Peace Research Institute (SIPRI) (1971–1975) into its study on chemical and biological warfare.

[2] There are numerous sources of information on the recommended procedures for the civil use of herbicides, for example, those by the US Agricultural Research Service (1969), Danielson *et al.* (1969), DeVaney (1968), Fryer & Evans (1968), Hilton *et al.* (1974), Klingman & Shaw (1971), J. Lawrence & Hollingsworth (1962–1969), Palm *et al.* (1968), Quisenberry (1970), Romancier (1965), R. Smith & Shaw (1966), and Spencer (1968). Bibliographies touching upon this subject have been prepared by Holm & Herberger (1971) and Kimmins & Fraker (1973).

The civil use of herbicides in the tropics has been covered by the American Chemical Society (1955), Bovey *et al.* (1968; 1969a), Brun *et al.* (1961), Commun (1961), Dowler & Tschirley (1970), Fryer (1974), Kasasian (1971), Little & Ivens (1965), Motooka *et al.* (1967), Richardson (1960), Schedler (1968), Steele (1968), Tschirley (1968), Wrigley (1961) and others.

[3] Operational aspects of the US chemical anti-plant spray programme have been described by the US Air Force (1971), US Army (1969), Darrow *et al.* (1969), Engineers (1972), Howard (1972:II), Irish *et al.* (1969), Young & Wolverton (1970), and others.

Application was made primarily (in about 95 per cent of all cases) from C-123 (UC-123) "Provider" transport aircraft (Fairchild Hiller Corp., Germantown, Maryland, USA) equipped with a 3 785-litre internal defoliant dispenser (Hayes International Corp., Birmingham, Alabama, USA, No. A/A 45Y-1) capable of delivering about 3 600 litres. Thirty-six high-pressure nozzles with an internal diameter of 9.5 mm are distributed on three booms. Droplet size was estimated at about 350 μm (mass median diameter). The equipment was pre-adjusted to deliver 28 litres/ha under standard operating procedures – an indicated air (flying) speed of 67 m/s and a height above vegetation of 46 m. Normal spray time was just over 2 min, although the entire payload could be ejected in about 30 s. One aircraft (sortie) treated a strip roughly 150 m wide and 8.7 km long, for an area of roughly 130 ha. A mission generally consisted of three to five aircraft flying in a staggered lateral formation. In order to keep off-target applications via drift and volatilization to an acceptable level, missions were not supposed to be flown when the wind speed exceeded 5 m/s or the ground temperature exceeded 29°C.

Helicopter application was carried out with the UH-1 "Huey" helicopter (Bell Aerospace Corp., Washington, D.C., USA) equipped with a 757-litre internal defoliant dispenser (Agavenco, Las Vegas, Nevada, USA), which was apparently only capable of delivering about 375 litres (meant to treat about 13 ha).

The following standard operating procedures for civil (domestic US) forest-spraying operations are provided as a means of comparison for the military ones noted above (Gratkowski, 1974): (*a*) use of low-volatile herbicides, (*b*) a drop size of 400 μm to 600 μm (mass median diameter), (*c*) air temperature less than 24°C, (*d*) relative humidity greater than 50 per cent, (*e*) indicated air (flying) speed less than 22 m/s, (*f*) wind speed less than 2.7 m/s and (*g*) height above vegetation as low as possible, preferably below 15 m to 30 m.

[4] There is a large literature on the ecological consequences of the use of herbicides, some of it polemical or uncritical. Important reviews for one reason or another are those by Carson (1962), Harper (1956–1957), House *et al.* (1967), Kuenen (1961), Mellanby (1970), N. Moore (1966; 1967), Mrak *et al.* (1969–1972), Rudd (1964), and Woodwell (1970). *See also* the bibliographies compiled by Condon (1968), G. Fox (1970), Headley & Erickson (1970), Ingram & Tarzwell (1954), Kimmins & Fraker (1973), and R. Thomas *et al.* (1964). Of further interest here is a series of ecological studies summarized by Young (1974) that was carried out over a period of years on a herbicide test site at Eglin Air Force Base in Florida, USA (*see also* Hunter & Young, 1972; Lehn *et al.*, 1970; Pate *et al.*, 1972; Valder, 1972).

Ecological assessments of the US chemical anti-plant programme in Indochina have been carried out *in loco* on behalf of the US Department of State in 1968 (Tschirley, 1969), the Society for Social Responsibility in Science in 1969 (Orians & Pfeiffer, 1970; Pfeiffer & Orians, 1972), the US Scientists' Committee on Chemical and Biological Warfare in 1969 (Westing, 1972a), the American Association for the Advancement of Science in 1970 (Constable & Meselson, 1971; Meselson *et al.*, 1971; Westing, 1971b; 1973a), the US Scientists' Institute for Public Information in 1971 (Pfeiffer, 1972; Westing, 1971–1972b), and the US Department of Defense between 1971 and 1973 (National Academy of Sciences, 1974). The several studies made were all severely circumscribed by the realities of the military and political situations, as all of the resulting reports have emphasized.

[5] The toxicity of herbicides to wild animal life has been reviewed or compiled by Condon (1968), Heath *et al.* (1972), N. Moore (1966), Pimental (1971), Tucker & Crabtree (1970) and Way (1969). Of interest here also is the veterinary toxicology literature dealing with herbicides. *See,* for example, Björklund & Erne (1966; 1971), Dalgaard-Mikkelsen & Poulsen (1962), Hassall (1965), Hilton *et al.* (1974), Palmer (1972), Palmer & Radeleff (1969), Radeleff (1970:VIII), Rowe & Hymas (1954), and Weimer *et al.* (1970). For effects of herbicides on fish, *see* note 7, below. For information on the poisonous herbicide contaminant dioxin, *see* note 9, page 44.

[6] The effects of herbicides on microorganisms (both bacteria and fungi) have been reviewed or compiled by Audus (1970), Bollen (1961), Condon (1968), Cullimore (1971), and others.

If all of a chemical anti-plant agent applied at military dose rates (table 3.1) were to reach the soil and become evenly distributed throughout the top 10 cm, and assuming that the soil weighs 2 650 kg/m^3 (Lutz & Chandler, 1946:236), then the following herbicide concentrations would result in the soil:

Agent Orange:	2,4,5-T: 6 mg/kg
	2,4-D: 5 mg/kg
	(dioxin: 26 ng/kg)
Agent White:	2,4-D: 3 mg/kg
	picloram: 1 mg/kg
Agent Blue:	dimethyl arsinic acid: 4 mg/kg
	(arsenic: 2 mg/kg)

The soil concentrations presented here are, of course, only crude approximations. However, three possible adjustments that might be made seem to more or less fully cancel each other out: an adjustment for the fraction not reaching the soil (x 0.5?), a second adjustment for downward penetration beyond 10 cm (x 0.5?), and a third adjustment for effective (functional) concentration in the soil water (x 4?).

The total concentrations given above will diminish more or less rapidly owing to volatilization, downward leaching and microorganismal and other decomposition. The effective (functional) concentrations will diminish to the extent that the herbicides become unavailable via adsorption (binding) onto soil particles.

[7] The effects of herbicides on aquatic plants have been considered by Gerking (1948), Gupta (1968), Hanson (1952), J. Lawrence & Hollingsworth (1962–1969), R. Smith & Shaw (1966) and others. The effects of herbicides on fishes and other aquatic animals (or on their sources of food) have been reviewed or compiled by Butler (1965), Condon

(1968), Cope (1966), Cope *et al.* (1970), Hanson (1952), Holden (1964), Ingram & Tarzwell (1954), Juntunen & Norris (1972), J. Lawrence & Hollingsworth (1962–1969), Mullison (1970), Pimental (1971), Sanders (1970), and others. For herbicidal effects on plants in general, *see* notes 1, 2 and 4, pages 41–42, and for animals in general, *see* note 5, page 43.

The fish of Indochina have been described several times (Kuronuma, 1961; Mizuno & Mori, 1970; Orsi, 1974). For information on the special coastal mangrove habitat, *see* note 10, page 45.

[8] The literature on the persistence and mechanisms of disappearance of herbicides in the soil and other components of an ecosystem is quite extensive, particularly as it applies to the phenoxy herbicides in the temperate zones. It has been reviewed or compiled by Aldrich (1953), Bovey & Scifres (1971), Hilton *et al.* (1974), Kearney & Helling (1969), Kearney *et al.* (1969a), Loos (1969), Norris (1970), Sheets & Harris (1965), R. Thomas *et al.* (1964), and others. In addition to the several papers referred to in the text that deal with tropical observations in general, those by Minarik & Darrow (1968) and by Blackman *et al.* (1974) are concerned specifically with South Viet-Nam.

A word might be added here regarding soil detoxification. Activated carbon has been suggested as a soil amendment or seed dressing for the detoxification of soils polluted with such hormonal herbicides as 2,4-D or 2,4,5-T (Bovey & Miller, 1969; Kearney *et al.*, 1969b; Lucas & Hamner, 1947; Orth, 1954; Schubert, 1967; Sheets & Harris, 1965). The application of chelating agents has been similarly suggested to deal with arsenic pollution (Batjer & Benson, 1958). The possibility of detoxifying aquatic freshwater ecosystems contaminated with 2,4-D via the introduction of certain green algae such as *Scenedesmus* (Scenedesmaceae) has been explored by Valentine & Bingham (1974) (*see also* Bingham, 1973).

[9] The Agent Orange used in Indochina contained relatively high levels of an exceedingly poisonous contaminant known as "dioxin" (sometimes as "dioxine" or as "TCDD") (2,3,7,8-tetrachlorodibenzo-*p*-dioxin). There is a burgeoning literature on this substance. Among the noteworthy publications are two compilations of papers (Blair, 1973; J. Moore, 1973), several review articles (Beek *et al.*, 1973; Crossland & Shea, 1973; Delvaux *et al.*, 1975; Epstein, 1970; Gribble, 1974; Helling *et al.*, 1973; Kimbrough, 1972; Lohs, 1973a; Neubert *et al.*, 1973), and an annotated bibliography containing 242 entries (Huff & Wassom, 1973).

The only measurements of dioxin concentration in the Agent Orange employed in Indochina that appear to be available are some that were made public by the US Air Force (1974:30). The Air Force reported that on its behalf the Dow Chemical Company (Midland, Michigan, USA) tested 200 samples of the agent chosen at random from among 26 689 available drums returned unused from South Viet-Nam after the employment of this agent had been discontinued in 1970 (thus a 0.75 per cent sample). The dioxin content of the 200 samples was reported to range up to a high of 60.4 g/m^3, and the arithmetic mean was given as 2.5 g/m^3, that is, 1.9 parts per million (ppm), by weight (*see* table 3.1).

It must be added here that even if the above reported 2.5 g/m^3 average dioxin contamination is accepted at its face value, this is almost certainly a very low figure to be used as an average for all of the Agent Orange expended in Indochina during the eight-year period of its use. The amount seems to be based almost entirely on samplings of agent manufactured toward the very end of the 1960s or even in early 1970. They were thus manufactured during a time when production methods had already been substantially improved over the early and mid-1960s when the bulk of the agent actually used in Indochina had been manufactured.

[10] General descriptions of particular note dealing with the mangrove (or "mangal") habitat and community have been prepared by J. Davis (1939–1940) and by Macnae (1968). Other publications which should be mentioned in the present connection for one reason or another include those by V. Chapman (1970), Golley *et al.* (1962), Heald & Odum (1970), Lugo & Snedaker (1974), McKinley (1957:63–71), and Walsh (1967).

Publications describing war damage to the mangrove community have been prepared by Constable & Meselson (1971), Cook (1970–1971), G. Davis (1973–1974; 1974), Desowitz *et al.* (1974), Drew (1974b), Lang *et al.* (1974:IV:91–125), Minarik & Bertram (1962), H. Odum *et al.* (1974), Orians & Pfeiffer (1970), Ross (1974), Sylva & Michel (1974), Tschirley (1969), Weatherspoon & Krusinger (1974a; 1974b), Westing (1971b), and others.

For the effects of herbicides on the aquatic habitat in general, *see* note 7, page 43.

Chapter 4. Mechanized landclearing

Where indicated thus,[1] the reader is referred to the notes on page 49.

I. *Introduction*

Forested areas can provide cover and sanctuary to military forces and agricultural areas can provide sustenance. Both are particularly important to guerrilla forces and in recognition of this, the United States during the Second Indochina War employed a variety of means to deny these advantages to its enemy. Discussed elsewhere are the employment for these purposes of chemical anti-plant agents (chapter 3) and high-explosive munitions (chapter 2). Discussed in the present chapter is the total removal of the offending vegetation using heavy mechanized landclearing equipment.[1] This is certainly the most straightforward approach to the problem, almost elegant in its conceptual simplicity.

The present chapter deals primarily with the pioneering employment by the USA of a particular type of heavy tractor equipped with a special blade, the so-called "Rome plough". A description of its use (section II) is followed by the ecological consequences of this use (section III). The more general and long-range ecological aspects of such actions are covered elsewhere (chapter 6), as are certain military considerations (page 83).

II. *Description*

The major instrument used by the United States in its landclearing operations in Indochina was a heavy tractor equipped with a large blade designed to sever and push aside (or first split, and then sever and push aside) trees of essentially any size.[2] Equipment of this sort has long been in use for civil landclearing operations both in the USA and elsewhere (Caterpillar Tractor Co., 1970).

In brief, clearing is accomplished by setting the sharpened lower edge of the angled blade to skim the soil surface. Brush, lianas and small trees provide no obstacle whatsoever to the forward motion of the tractor whereas large trees are first split once or perhaps twice with the blade's lance ("stinger") prior to being dispatched. In practice, a group of perhaps two dozen tractors equipped with Rome-plough blades plus several equipped with bull blades function in concert.[3] Depending upon local conditions, the debris is left in place to rot, is windrowed and burned, or is shoved into gullies and craters. Actually, the debris provided the military with not much of a disposal problem in South Viet-Nam whenever the local inhabitants were permitted to salvage the wood. The character of the colonizing vegetation and the military situation usually determined whether at some future date a cleared area was sprayed with herbicides or perhaps rescraped.

The USA began its landclearing operations in Indochina on a small scale in 1966 using tractors equipped in some instances with bull blades and in others with Rome-plough blades (Canatsey, 1968). The early efforts were devoted largely to the clearing of roadsides and along other lines of communication (LOCs) for the purpose of discouraging enemy ambushes. By the beginning of 1968 most of the major road systems in Military Regions II and III had been cleared to a distance of 100–200 m and occasionally even 300 m on each side. These endless swaths throughout forest and plantation are now a conspicuous feature of the regional landscape.

In mid-1967 some of the tractors were organized on a trial basis into small units for the purpose of clearing small woodland tracts. It soon became evident that the total removal of a forest as a means of area denial to guerrillas was in some ways far superior to such alternate techniques as aerial herbicide spraying or interdiction shelling and bombing. By mid-1968 massed tractors organized into companies were increasingly employed to scrape bare large contiguous tracts of wild land.[3]

All told, some 325 x 10^3 ha were cleared by the Rome ploughs in South Viet-Nam, that is, 2 per cent of its entire land area (US Army, private letter, 16 October 1973). No breakdown of this total has been released, either as to year or Military Region. At least some Rome ploughing occurred in all of South Viet-Nam's Military Regions, and even some in the Fish Hook region of adjacent Cambodia (Massey, 1970; Ploger, 1974:176). However, the greatest activity occurred in Military Region III. Obliterated there, for example, were the 7 100-ha Hô Bô woods in central Bien Hoa province, the 3 600-ha woods of the same name in west-central Binh Duong province and the 2 700-ha Boi Loi woods in southeastern Tay Ninh province. In Military Region I, the Batangan peninsula in Quang Ngai province was subjected to extensive landclearing.

It can be added that US forces experimented with a number of additional approaches to mechanized landclearing. For example, rubber (*Hevea brasiliensis*, Euphorbiaceae) plantations as well as young forest stands were occasionally knocked down by dragging a heavy chain stretched between two tractors through such stands (R. Adams, 1971). Satisfactory results were apparently achieved using a chain between 45 m and 75 m in length and weighing 60 kg/m (*but see* Hay, 1974:87; Ploger, 1974:97–98). Finally a giant device known as a "tree crusher" (Sterle, 1967) was employed on a trial basis, but proved to be unsatisfactory in the Indochina theatre (Hay, 1974:87); Ploger, 1974:98–99).

III. *Ecological consequences*

The environmental impact of mechanized landclearing can take on serious dimensions, particularly when thousands of contiguous hectares are abused in this fashion. The magnitude of the impact is, of course, strongly influenced by a number of factors, making generalizations difficult. The most important of these variables are: (*a*) the absolute size of the area treated and whether or not it is in one conterminous parcel, (*b*) the size of the area relative to the size and vegetational status of the overall region, (*c*) the thoroughness of treatment, and (*d*) site factors such as the degree of slope, soil type and rainfall patterns. In the longer term one must add to this list the character of the pioneer biotic community that subsequently becomes established on the area as well as any subsequent human manipulations of the area, both military and civil.

It is evident after an area has been scraped bare that the first major concern is soil erosion. The damage to the site from erosion can be massive during the months before a new vegetational cover becomes established, especially in hilly terrain and in terrain with heavy rainfall. A number of official US statements to the contrary, much if not most of the soil surface is exposed and disturbed by the Rome-plough clearing operations. For example, two US Army landclearing engineers in Military Region I observed that during Rome-ploughing operations, "the smell of damp, freshly turned earth laced with diesel fumes pervades the air" (Mattingly & McVann, 1971:6). The landclearing operations also bring about additional, more subtle soil degradation via the loss in solution of mineral nutrients, the so-called nutrient-dumping phenomenon which is described in some detail elsewhere (*see* page 66). Moreover, there is an aggravation of flooding problems prior to the establishment of a new plant cover.

The exceedingly rich tropical faunal community that depends for its food and shelter on the forest cover being eliminated by mechanized landclearing is, of course, decimated in the process. To quote again from the first-hand account of the two army engineers who were describing their landclearing experiences in Military Region I: "Deer [Cervidae], wild pigs [*Sus*, Suidae], and pheasants [Phasianidae] are often seen fleeing from the area being cleared as [it] shrinks under the relentless bites of the dozers. Snakes up to ten feet [3 m] long have difficulty running from the dozers" (Mattingly & McVann, 1971:6). To this must be added that the reprieve for most of the surviving animals is probably rather short-lived since the unaffected areas to which they might escape are likely already to be populated to their maximum carrying capacity. Nor can these animals somehow wait for the new plant community that will eventually recolonize the area. This new community will provide a remarkably different and completely unsuitable habitat for most of the original animals.

Mechanized landclearing operations lead to a remarkable alteration of the microclimate, that is of the climate in the region of the interface between atmosphere and geosphere. The changes which occur include elevated insolation, raised temperatures, increased wind velocities, decreased humidity and reduced levels of carbon dioxide. The ecological consequences of such changes are discussed elsewhere (*see* page 74).

During the months following landclearing operations, the complex original forest community is replaced by an invading pioneer community. This new community will have a far more simple floristic composition than the original one. It will often be dominated by herbaceous or woody grasses (Gramineae) and only rarely by woody dicotyledons such as *Dipterocarpus* (Dipterocarpaceae). Many years can be expected to elapse before natural forces will restore a Rome-ploughed site to its former status.

IV. *Conclusion*

There is no doubt that Rome-plough-equipped tractors will remain in the array of heavy equipment employed by military engineer units. Certainly they will continue to be used for clearing vegetation from roadsides and other lines of communication as well as in base camp areas and around their perimeters. These relatively modest uses should provide little cause for ecological concern.

What does provide cause for concern about the potential future of landclearing tractors is their utility in large-scale area-denial operations of the sort pioneered in Indochina. The employment of the so-called Rome ploughs as counterinsurgency weapons in forested regions has been widely acclaimed by the military as one of the outstanding innovations of the Second Indochina War (*see* page 84).

It thus appears that vast land areas may be cleared with tractors in counter-insurgency and other wars of the future, with potentially devastating results to the local ecology.

Notes

[1] Mechanized landclearing with "Rome ploughs" and other heavy equipment has been described or evaluated by a number of military authorities (Canatsey, 1968; Draper, 1971; Engineers, 1972; Fabian, 1970; Hay, 1974:87–89; Howard, 1972; Kerver, 1974; Kiernan, 1967; Ploger, 1968; 1974:95–104; Rogers, 1974:61–66; Savage, 1973) and others (Haseltine & Westing, 1971; Westing, 1971d). *See also* page 84.

[2] The basic tool of the US landclearing operations in Indochina was a heavy tractor equipped with a special landclearing blade. The tractor is manufactured by the Caterpillar Tractor Co. (Peoria, Illinois, USA), and is its Model D7E. The clearing blade is manufactured by the Rome Plow Co. (Cedartown, Georgia, USA), its Model KGA7E. Various safety features including armour plate were added. The total assembly weighed approximately 33 x 10^3 kg. The tractor weighs about 18.1 x 10^3 kg, the blade 2.2 x 10^3 kg, and the added armour 12.7 x 10^3 kg.

The adjustable blade is 340 cm wide and has a projecting "stinger" tip on the left side about 105 cm long. The horizontally oriented lower cutting edge of the blade as well as the vertical stinger must be kept sharpened (for which a portable grinder is generally used). The angle mounting of the blade results in a shearing action which is capable of severing at ground level all but very large trees. The latter are first split by the stinger. The guide bar mounted above the blade windrows the debris to the right of the tractor.

A much larger machine – the Caterpillar D9 tractor equipped with the Rome KGA9 blade – was used by the USA in South Viet-Nam on a trial basis. The complete assembly weighs more than twice the one above and the blade is just over 4 m wide. However, this machine was not considered a success in Indochina for a variety of reasons.

[3] The US Army Corps of Engineers organized its landclearing tractors in Indochina into about half a dozen companies, each consisting of three tractor platoons plus one maintenance platoon. The landclearing platoons each contained ten tractors, nine equipped with clearing blades and one with a bull blade.

When clearing forest land, an area of perhaps 50 ha at a time would usually be designated for a tractor company to level. The two dozen or more functioning tractors, operating in a drawn-out, staggered formation, then worked around and around the periphery, slowly closing in on the centre in ever smaller counterclockwise circles until nothing was left. Generally the lead tractor was directed for much of the time by the company commander circling the cut overhead in a small helicopter.

Early in the programme it was determined that tractors equipped with clearing blades could, under tactical conditions, dispose of jungle at the rate of 0.5 ha/h whereas a similar tractor equipped with a bull blade could only progress at the rate of 0.2 ha/h (Canatsey, 1968). Subsequently it was estimated that with an average of 25 tractors operational at any one time, a company could clear 160 ha/d of light jungle, but only 40 ha/d of heavy jungle (Engineering News-Record, 1970). For medium jungle, the figure ranges from 60 ha/d to 80 ha/d (Ploger, 1974:102).

Chapter 5. Miscellaneous weapons and techniques

Where indicated thus,[1] *the reader is referred to the notes on page 61.*

I. *Introduction*

Previous chapters have been devoted to examining the ecological impacts of high-explosive munitions (chapter 2), chemical anti-plant agents (chapter 3), and landclearing tractors (chapter 4) as these were used during the Second Indochina War. A number of weapons and techniques of potential ecological concern also used during this war remain to be examined. These include blast munitions (section II), anti-personnel chemicals (section III), weather manipulation (section IV), flooding (section V) and the use of fire (section VI). These miscellaneous approaches to counterinsurgency warfare are all capable of causing environmental damage, at least under special conditions. For example, some could become cause for serious concern if extensively employed, others if the site conditions (climate, and so on) were appropriate, and still others if the techniques were "improved" somewhat.

II. *Blast munitions*

Discussed in the present section are two special-purpose bombs which were used to a limited extent during the Second Indochina War. Both of these types of bombs depend for their primary effectiveness upon the blast (shock) wave they create, as opposed to the general-purpose high-explosive bombs that depend for their effectiveness primarily on the flying metal fragments ("shrapnel") they produce (chapter 2). One of these new munitions employs a slurry explosive *in lieu* of the more usual 2,4,6-trinitrotoluene (TNT) preparation and the other a volatile organic liquid. Although neither was used to any major extent in Indochina, they deserve at least brief examination because of the wide-area impact per individual bomb.

Beginning in 1967, the USA during the course of the war expended more than a hundred large concussion bombs containing slurry explosive, most of them in South Viet-Nam but some also in Cambodia and Laos. The major purpose of dropping these concussion bombs was to create an instant helicopter landing zone even in dense jungle, but they were occasionally also used for other purposes (anti-personnel, anti-matériel, and so on). By far the most commonly used size, Model BLU-82/B, was a huge bomb weighing almost 7×10^3 kg.[1]

These concussion bombs were dropped from specially equipped C-130E "Hercules" transport aircraft (Lockheed Aircraft Corp., Burbank, Cal., USA) from altitudes of between 2 and 3 km, and occasionally even from over 6 km, their free fall arrested somewhat with a small drogue parachute. The bomb is constructed so as to detonate about 1 m above the ground, the charge at that time being set off simultaneously at both ends of the bomb. The resulting horizontal radial blast (about twice that of the vertical blast) leaves no crater. Rather it uproots, shears, shatters and blows away all trees and other obstructions to create a virtually perfect clearing even in heavy jungle. The instant openings which were in this way blasted out of the forests of Indochina had an average diameter of 130 m and thus an average area of 1.3 ha (US Air Force release, 10 January 1972).

The 1.3 ha zone of total forest clearing just mentioned defines, of course, the zone per concussion bomb of total vegetational obliteration. Beyond this zone there will be an additional one of partial damage, although its size is difficult to estimate.

Regarding the impact on wildlife, one can turn to the available human casualty data. According to the manufacturer, the bomb's lethal range (presumably the range of lethality for 50 per cent of exposed individuals, or LD_{50}) is 74 m, a distance that defines an area of 1.7 ha (IRECO, Salt Lake City, Utah, USA, private letter, 20 March 1972). Moreover, according to a US Air Force release (10 January 1972), the total casualty zone for this bomb extends outward for 396 m and thus includes an area of 49 ha.

The second category of blast munition to be discussed in this section is the so-called fuel/air explosive bomb.[2] The use of these bombs was also begun by the United States in 1967, but, unlike the concussion bombs, their use remained experimental. A variety of styles and sizes were tried out, with a total of at least 80 having been dropped at one time or another. Fuel/air explosive bombs were under development primarily for the purpose of setting off emplaced pressure-sensitive land mines and booby traps, but were also tried against personnel and matériel.

The free fall of most of the fuel/air explosive bombs used in Indochina was arrested with a small drogue parachute. After the bombs reach the ground and burst, there is a very brief delay (of 125 ms or so) before detonation occurs, which permits an appropriate spreading and mixing of fuel with air. The resulting explosion and blast wave also produce no crater. A miscellany of information having ecological relevance follows.

According to Robinson (1973), a 227-kg fuel/air explosive bomb containing 98 kg of fuel is designed to spread a blanket of vaporized fuel about 2.4 m thick over a circle having a diameter of 15.2 m, and thus an area of 0.02 ha. Then, upon detonation, this entire zone is reported to be subjected to a transient overpressure (that is, pressure above atmospheric) in excess of 2000 kPa. On the basis of a published photograph (Cannon, 1971:136), it can be estimated that a 1 134-kg fuel/air explosive bomb produces a forest clearing of roughly 1 ha in size (that is, with a diameter of roughly 113 m). Although the picture depicts quite a few bare stems still standing within the clearing, these are presumably dead. The zone in which a 250-kg fuel/air explosive bomb produces a 50 per cent probability of human casualties has been said to fall between 1 and 10 ha (Red Cross, 1973:26). It is additionally important to mention that when a fuel/air explosive bomb malfunctions in some ways, it can produce a fire-ball rather than a blast wave (Björnerstedt *et al.*, 1973:16), thereby creating the possibility of initiating a wild fire (*see* section VI).

Thus it can be seen that in terms of environmental consequences, the two categories of blast weapon described in this section simply destroy all living things – both plant and animal – over an area of the order of one hectare or so in size. Beyond this inner circle of annihilation, there exists a further zone of indiscriminate death and serious injury to wildlife that may cover an area of up to several hectares or even tens of hectares in extent. Any significant injury to wildlife is, of course, tantamount to death. It should thus be apparent that the overall ecological impact of blast munitions will depend largely upon the intensity of their employment.

III. *Anti-personnel chemicals*

The chemical anti-personnel agent "CS" (*o*-chlorobenzalmalononitrile) was used in large amounts and in various ways by the United States during the Second Indochina War. Since this action represents a significant chemical intrusion of the environment, it must thus be examined from the standpoint of ecological impact.[3]

CS results in militarily significant harassment of unprotected personnel at a particulate aerosol concentration in the atmosphere somewhat above 1 mg/m^3. Used at this level, CS produces intense lacrimation (crying), sternutation (sneezing), and irritation of the upper respiratory tract. However, a tactical innovation of the Second Indochina War of particular importance in the present context was the employment of CS1 (a finely pulverized form of CS) for protracted area denial (Blumenfeld & Meselson, 1971). An especially non-degradable form – CS2 – was developed in 1968 to facilitate this mission. Whereas the application of CS1 can render an area inhospitable for perhaps 15 days (Army, 1969:16), that of CS2 will do so for 30 to 45 days (Cannon, 1971:146). Precise durations depend, of course, on such factors as initial level of application and subsequent rainfall.

Although the US Department of Defense has released no information on US expenditures of CS in Indochina, one can at least gain an indication of their magnitude from the procurement figures available for the war years (table 5.1). The total quantity of CS overtly procured by the US Department of Defense during this time amounted to about 9×10^6 kg, about four-fifths of this in bulk form and the remainder directly incorporated into munitions. Perhaps 30 per cent of the total was in the more persistent CS2 form. The level of CS application considered necessary by the military to achieve satisfactory area interdiction appears to have fallen somewhere between 1 kg/ha and 10 kg/ha. It is thus evident that procurement was sufficient to interdict at one time or another

Table 5.1. US procurement of CS gas during the Second Indochina War: a breakdown by type and year

tonnes = 10^3 kg

Fiscal year	CS in bulk	CS1 in bulk	CS2 in bulk	CS in munitions	CS1 in munitions	Total
1961–62	. .	. .	–	. .	. .	**. .**
1962–63	. .	. .	–	. .	. .	**. .**
1963–64	102	64	–	106	. .	**272**
1964–65	42	83	–	42	. .	**167**
1965–66	171	552	–	–	4	**727**
1966–67	198	349	–	797	26	**1 371**
1967–68	324	1 474	131	350	6	**2 284**
1968–69	915	73	1 762	351	26	**3 127**
1969–70	–	161	830	58	13	**1 062**
1970–71	–	–	–	33	8	**41**
1971–72	. .	. .	. .	. .	. .	**. .**
1972–73	. .	. .	. .	. .	. .	**. .**
Total	**1 753**	**2 755**	**2 723**	**1 737**	**83**	**9 052**

Notes and sources:

(*a*) Where possible, the above data are derived from those of Fulbright (1972:307). Values not obtainable from this source are derived largely from those of Mahon (1969:124), whereas those not obtainable from either of these sources are derived from those of McCarthy (1969:15765). Amounts of CS contained in the munitions are from the US Army (1969:II). Those not obtainable from this source are derived to the extent possible from the US Army (1967b), whereas those not obtainable from either of these sources are from the US Army (private letter, 17 June 1974).

(*b*) Missing data have not been released by the US Department of Defense.

(*c*) Information on CS expenditures in Indochina has never been released by the US Department of Defense.

(*d*) CS is the code name for *o*-chlorobenzalmalononitrile. CS1 is a finely pulverized (micronized) form of CS, whereas CS2 is a CS powder that has been treated to make it water-repellent and thus less rapidly degradable under field conditions.

during the course of the war between 1 x10^6 ha and 9 x 10^6 ha of Indochina. Although at least some CS was employed against all four of the Indochinese countries, most of that used was expended in South Viet-Nam with its 17 x 10^6 ha.

It appears reasonable to assume that the CS so liberally applied to the South Viet-Namese environment had no major ecological effects inasmuch as none has been reported in the literature. Any more subtle effects could, of course, have escaped attention. These would hinge upon the toxicity of CS to the various biota exposed to it during the several weeks of the chemical's existence following its field application. The modest amount of pertinent information available on the subject follows.

To begin with, it appears that CS harasses and is toxic to the warm-blooded vertebrates at levels roughly comparable to those for man. For example, Punte *et al.* (1962) found that the rats (*Rattus rattus*, Muridae), mice (*Mus musculus*, Muridae) and pigeons (*Columba livia*, Columbidae) were only slightly more resistant than man to CS via inhalation, whereas guinea pigs (*Cavia porcellus*, Caviidae) and rabbits (*Oryctolagus cuniculus*, Leporidae) were somewhat more sensitive. It seems, moreover, that chickens (*Gallus gallus*, Phasianidae) are somewhat less irritated by a given level of CS aerosol exposure than are humans.

With respect to aquatic biota, Ward (1973) has reported that the common killifish (*Fundulus heteroclitus*, Cyprinodontidae) is killed by 4 g/m^3 in the ambient water (50 per cent mortality within 96 hours, that is, 96 h LC_{50}), and that the duckweed *Wolffia papulifera* (Lemnaceae) is injured by concentrations of 5 g/m^3 and killed by 100 g/m^3. At least the mammalian toxicity of CS is partially attributable to its *in vivo* conversion to cyanide (Frankenberg & Sörbo, 1973).

CS appears also to be somewhat toxic to terrestrial vegetation, injury to trees having been reported following their exposure during a civil riot-control operation (Cockrell, 1971). In this regard it can also be noted that one compound or another chemically related to CS has been reported to have herbicidal, fungicidal, insecticidal, or nematocidal properties (Jones, 1972), attributes thus possibly shared to a greater or lesser extent by CS.

It seems safe to conclude that military field applications of CS result in at least modest and transient ecological perturbations.

IV. *Weather manipulation*

Military activities can presumably modify meteorological phenomena either inadvertently or by conscious manipulation. The former possibility is described elsewhere (*see* page 74), whereas the latter is the subject of the present section. Indeed, although it did not become generally known until the latter part of the war, the United States carried out rather extensive attempts to manipulate the rainfall in Laos and elsewhere in Indochina.[4]

At least between 1966 and 1972 the USA annually seeded cumulus clouds over Indochina (table 5.2). The seeding agents employed for this purpose included silver iodide and lead iodide. The major reason for these efforts was reported to be interdiction of enemy lines of communication, particularly the supply routes in southeastern Laos. It was attempted each year to intensify and prolong the annual rainy season in order to make the so-called Ho Chi Minh trail sufficiently muddy to render it impassable, or at least more difficult to use.

In North Viet-Nam, cloud seeding (through the use of undisclosed chemicals) may have been carried out largely to make inoperable (to "attenuate") the enemy radars used for aiming defensive surface-to-air missiles (Hersh, 1972). Other reported uses in Indochina have included the production of weather sufficiently bad to hamper enemy offensives, the altering of rainfall patterns to aid US bombing missions, the providing of inclement weather to enable the success of covert ground operations, the creation of generally disruptive floods and the diversion of enemy manpower to undoing the mischief caused by the bad weather instigated.

Although the military seemed satisfied with the level of success of its weather-modification operations in Indochina (Gravel *et al.*, 1971–1972: IV:421; Pell, 1974:103–108), a dispassionate arbiter would be hard put to recognize the basis for this optimism.

Assuming that the military operations did, in fact, materially augment the regional rainfall in Indochina, then one can speculate on a number of possible consequences. With rainfall augmentation there is, first of all, the possibility of increased flood damage (*see* page 73). It remains unknown for example, whether the serious flooding that occurred in North Viet-Nam in 1971 (Darcourt, 1971; Vietnam Newsletter, 1971) can be attributed at least in part to US weather manipulations. Indeed, 1971 was the peak year of US cloud-seeding activity (table 5.2), and even if this was carried out only in Laos, as announced, the prevailing winds during the rainy (cloud-seeding) season are southwesterly, that is, from Laos to North Viet-Nam. Secondly, with increased rainfall there is an aggravation of erosional damage, particularly in hilly terrain and especially in terrain previously disrupted by bombing or other military activity (*see* page 65).

Water-dependent insects, including disease vectors, would be favoured in times of increased rainfall, with possible subsequent increases in the incidence of various diseases of wildlife, livestock and human beings. Moreover, there might have been instigated a variety of more or less subtle imbalances in the ecosystems being thus tampered with, including modest changes in reproductive, growth, or mortality rates of some species. The resultant ecological debilitations would not become apparent without detailed examination.

The cloud-seeding agents themselves – silver iodide, lead iodide, and so on – exert at least a minor adverse effect on the ecosystems into which they are introduced. Certain aquatic biota such as algae, invertebrates, and some fish appear to be the most vulnerable to such abuse (Cooper & Jolly, 1970).

In conclusion it is useful to note that the several published analyses of the potential ecological impact of widespread weather modification for civil purposes all agree that the ultimate outcome of even such well-meant activities is if anything likely to be undesirable.[4] Seemingly minor changes in insolation, temperature or precipitation can bring about substantial and often unexpected changes in the affected ecosystems.

V. *Flooding*

Under specialized geographical conditions it is possible by appropriate military actions to bring about flooding of an area. For example, one of the ramifications

Table 5.2. US cloud seeding operations in the Second Indochina War: a breakdown by year and region

Year	Seeding cartridges expended					
	South Viet-Nam	North Viet-Nam	Cambodia	Laos	Total	Total sorties flown
1961	–	–	–	–	–	–
1962	–	–	–	–	–	–
1963	(Several)	–	–	–	(Several)	(Several)
1964	–	–	–	–	–	–
1965	–	–	–	–	–	–
1966	–	–	–	(560)	(560)	56
1967	(Several)	1 017	–	5 553	6 570	591
1968	–	98	–	7 322	7 420	(734)
1969	–	–	–	9 457	9 457	528
1970	–	–	–	8 312	8 312	277
1971	–	–	–	11 288	11 288	333
1972	(1 000)	–	(1 000)	(2 362)	4 362	139
1973	–	–	–	–	–	–
Total	**(1 000)**	**1 115**	**(1 000)**	**(44 854)**	**(47 969)**	**2 658**

Notes and sources:

(*a*) The data are from Pell (1974:92–102) except for the 1963 information which is from Hersh (1972) and Shapley (1974).

(*b*) The seeding cartridges generally used generated either a silver iodide or lead iodide aerosol (Pell, 1974:91). The devices may have been similar to the commercially available "Weathercord" (Weather Engineering Corp., Dorval, Quebec, Canada) which contain 518 mg of silver iodide (Goyer *et al.*, 1966).

of augmenting the rainfall in a region via military cloud seeding might be the instigation or aggravation of flooding. Moreover, the likelihood exists of the incidental aggravation of flooding in conjunction with large-scale military land-clearing operations via, for example, chemical anti-plant agents or Rome ploughs (*see* page 74).

Where the situation lends itself to this, the most straightforward means of causing flooding is to destroy dams or dikes by bombing or other means. Asian enemies of the United States, both existing and potential, were explicitly warned during the 1950s by word and deed of their vulnerability in this regard (Air University Quarterly Review, 1953–1954; Rees, 1964:381–82). Indeed, the destruction of irrigation dams was considered by the USA to be among the most successful of its air operations of the Korean War (Futrell *et al.*, 1961:627–28, 637).

During the Second Indochina War, the United States again attacked agriculturally important dams, dikes and sea-walls through bombing and shelling, especially in North Viet-Nam. Numerous examples of such damage have been reported (Duffett, 1968:226–35; Lacoste, 1972; Westing, 1973b). During this war, however, such damage was, according to the US official view, either inadvertent or "collateral" (Gliedman, 1972; Porter, 1972).

Be that as it may, flooding can have a significantly adverse impact on the local or regional ecology by drowning flora and fauna, by accelerating erosion (page 64), by contaminating the soil with salt when seawater is involved (Dorsman, 1947) and in other ways (*see* page 73).

VI. *Fire*

Fire has been used in war since ancient times. Although the Second Indochina War was no exception, intentionally set rural wild fires for purposes of area denial or similar widespread harassment played only an indifferent role in this conflict.[5] Actually, incendiary weapons such as magnesium-encased thermit bombs and grenades, white phosphorus bombs and shells, and napalm bombs and canisters were employed during the Second Indochina War in quantities that exceeded those of any prior war in history (SIPRI, 1975:I). Their use, however, was confined in large part to close-air-support missions. The several known instances of attempted wide-area forest destruction are examined in the present section.

What seems to have been the militarily most successful wide-area incendiary attack of the war was carried out in the U Minh forest, a US enemy stronghold in Military Region IV. For several weeks during the spring of 1968 the United States was able via repeated heavy incendiary attacks to nurture there some fires of uncertain origin (Time, 1968). In another instance, the United States was reported in the spring of 1971 to have dropped enormous quantities of incendiary devices onto the forests around a besieged outpost in west central Kontum province in Military Region II (Associated Press, 1971b). No evaluation of either of these attacks appears to be available.

The most noteworthy instances of rural incendiary attack in the present context were three major attempts by the USA between 1965 and 1967 to initiate massive forest fires over extensive enemy-controlled areas (Hartmann, 1967; McConnell, 1969–1970; Perry, 1968; Randal, 1967; Reinhold, 1972; Shapley, 1972c). In each of these cases a large forest area was prepared by prior herbicide spraying in order to provide an adequate fuel base of dead leaves and twigs (McConnell, 1969–1970). The first of these attempts, "Operation Sherwood Forest", was carried out during the spring of 1965 for the purpose of destroying the almost 3×10^3 ha Boi Loi woods in southeastern Tay Ninh province in War Zone C. During "Operation Hot Tip" in early 1966 an attempt was made to destroy perhaps 7×10^3 ha of forest in the Chu Pong mountains in north central Pleiku province (Military Region II). And in "Operation Pink Rose" early in 1967 attempts were made to destroy almost 8×10^3 ha of forest near Xuan Loc in south central Long Khanh province (War Zone D).

None of these three carefully planned and executed attempts at initiating self-propagating wild fires was successful despite the herbicidal pretreatments, the massive use of incendiary devices, and the sundry technical refinements that were added from one operation to the next. The failures can, of course, be attributed to the generally wet conditions (high humidity and/or rainfall) that prevail in the region. The finely divided fuels necessary to sustain a forest fire take up too much moisture to support combustion when the ambient relative humidity is above 80 per cent, the usual condition that obtains throughout much of Indochina. Additionally, not much litter accumulates on the tropical forest floor as a potential source of fuel, and Mutch (1970) suggests that tropical rain forests provide a relatively poor grade of fuel even if permitted to become dry (*see also* Batchelder & Hirt, 1966). However, inasmuch as serious attempts were made during the Second Indochina War to initiate large-scale self-propagating wild fires, it becomes important to evaluate such actions from an ecological standpoint.

Fire has always been a natural factor more or less importantly involved in the shaping of terrestrial ecosystems. Since the beginning of time there have been innumerable fires caused by lightning. And wherever man has lived, from antiquity to the present day, he has continued to start fires, either by accident or design. For a variety of meteorological and other reasons, the frequency and severity of wild fires differ from region to region, running the gamut from common and extensive to virtually non-existent. In the regions of relatively high frequency, fire is, in fact, the dominant factor determining vegetational (and thus faunal) composition (Ahlgren & Ahlgren, 1960; Cooper, 1961). This holds true, irrespective of latitude. Thus, Lutz (1956) has shown that the vegetation characteristic of interior Alaska is determined by the periodic fires that have always been common to that region. Certain forms of grasslands have become established and perpetuated only through the action of repeated fires, both in temperate zones (Sauer, 1950; Wells, 1965) and in the tropics (Budowski, 1956; Holmes, 1951; Wharton, 1966; 1968).

One might suggest that, as a general rule, the ease with which wild fires can be started and sustained in any particular region is directly correlated with the ability of the natural local ecosystems to survive such assault. The plants indigenous to fire-prone regions have presumably evolved fire-survival mechanisms and, in some instances, have even come to depend upon periodic fires in one way or another. Conversely, those plants found in a region rarely subjected to fire could be expected to be decimated by a conflagration.

For the many regions throughout the world where natural wild fires are an occasional occurrence, a forest fire is relatively easy to set (at least at certain seasons of the year) and will result in more or less severe ecological damage. In such a region a forest fire will injure many of the large trees and kill some of them. The degree of initial damage depends upon the weather conditions at the time of the fire and upon the species involved. The ability of a species to withstand fire damage is for all but young individuals to a great extent a function of its bark characteristics, mainly the thickness. Seedlings and saplings are, of course, more highly susceptible to initial damage than larger trees. However, the greatest amount of damage to the trees is caused by the fungi which gain entry through fire wounds (Hawley & Stickel, 1948:III), and to a lesser extent by the insects which do so (J. Hodges & Pickard, 1971).

Major forest fires can also do a certain amount of harm to the ecosystem via damage to the soil (*see* page 65). An important source of such damage is through a reduction in the amount of soil litter (that is, of the soil's A_0 horizon). Some of it will burn in the fire, but the greatest loss is via a subsequent diminution of the required continuing replenishment. Litter with its associated humus is a particularly scarce commodity in the tropics.

With a reduced layer of protective litter, the soil becomes subject to increased erosional damage and associated problems (Arend, 1941; Lowdermilk, 1930). The loss of nutrients in solution, or nutrient dumping, will be a particular problem, perhaps especially for soluble phosphorus (McColl & Grigal, 1975). Moreover, flood danger will be increased.

Fires can also raise havoc with the wildlife in an area, both directly and indirectly, although the degree of damage to any particular species can differ markedly depending upon the season of the year. The direct faunal destruction that is likely to occur as a result of a wild fire has been vividly described by D. Kipp (1931). Fires are, of course, harmful to animals indirectly via destruction of their food and cover. Moreover, the very different habitat that is likely to develop following a major fire, that is, following vegetational recolonization, will support a new and far less diversified animal community.

It can be concluded that the military fires attempted in Indochina were of only minor ecological import owing to the regional site conditions. However, it must be noted that large rural military fires are likely to be more damaging than seemingly comparable rural fires of non-military origin (even those maliciously started). Not only will the military fires be likely to have been set via the massive and repeated applications of efficient incendiary devices, but efforts to extinguish them are likely to be hampered by concomitant military actions of various sorts. Moreover, if an area not normally subject to natural fires can be ignited by improved incendiary techniques, including appropriate pretreatment, a technology under development (Bentley *et al.*, 1971; Forman & Longacre, 1969–1970; Philpot & Mutch, 1968), then the ecological impact could be a serious one.

VII. *Conclusion*

Several disparate weapons and techniques have been briefly examined in the present chapter. They have in common the potential of being able to debilitate the environment under a variety of special circumstances. As a result, these techniques of war demand special ecological consideration whenever their use is contemplated.

Since this is a chapter of military miscellany, it may not be going too far astray to note that the non-hostile military activities in a theatre of war also bear continued scrutiny with respect to their incidental environmental impact. For example, the ecological impact of widespread military landclearing operations has been discussed elsewhere (*see* page 46). However, additional thousands of hectares of forest land were cleared particularly in South Viet-Nam in conjunction with constructing base camps, airfields, highways and so forth. Although not further considered here, the ecological impact of such activities may not have been insignificant. Another case in point was the introduction into the Indochinese environment of malathion (S-(1,2-dicarbethoxyethyl)-O,O-dimethyldithiophosphate). Millions of hectares of South Viet-Nam were aerially treated with this insecticide over the years as a part of the US programme of malaria control, with unknown ecological side effects.

Notes

[1] Not much information has been released on the BLU-82/B and similar concussion bombs. The US Air Force has released a very brief (72 s) film showing the bomb in action (Newsfilm Release No. 101–70, 19 June 1970) and also a two-page "fact sheet" (10 January 1972). One can also refer to a number of reports in the press (Foisie, 1970; Life Magazine, 1971; Whitney, 1971) and to a description of the bomb and its use in Indochina (Westing, 1972c).

The BLU-82/B concussion bomb has the following dimensions: total weight 6 804 kg, explosive content 5 715 kg, diameter 137 cm, length 345 cm, length of fuse probe 97 cm, thickness of steel case of main cylinder 6 mm, and thickness of steel case of nose cone 13 mm. The explosive is a gelled aqueous slurry of ammonium nitrate and aluminium powder (designated as DBA-22M). This explosive has an energy yield (its so-called heat of explosion) of roughly 9 MJ/kg, that is, about twice that of 2,4,6-trinitrotoluene (TNT).

Although data have not been released on the duration of the positive phase (or of the rise time) of the blast wave of the BLU-82/B bomb, the following blast wave parameters may still be of some interest. They have been calculated from bomb and explosive specifications supplied by the manufacturer (IRECO, Salt Lake City, Utah, USA, private letter, 20 March 1972). At a radius of 65 m (the average zone of complete clearing) the total (Mach front) peak transient overpressure is 455 kPa whereas the associated peak transient wind (particle) velocity is 497 m/s. Comparable values for a radius of 74 m (the so-called lethal zone) are 345 kPa and 420 m/s. Comparable values for a radius of 396 m (defining the casualty zone) are 17 kPa and 38 m/s.

[2] As with the concussion bombs, there has not been much information released on fuel/air explosive (FAE or FAX) bombs. The most informative article is by Robinson (1973) and there are a number of other available accounts (Associated Press, 1971a; Björnerstedt *et al.*, 1973:16; Cannon, 1971:135–36; National Defense, 1973–1974; Ordnance, 1971–1972; Red Cross, 1973:26, 40; Wulff *et al.*, 1973:37–38).

Fuel/air explosive bombs have been manufactured in several sizes, including one weighing 227 kg and another 1 134 kg. Some models consist of one large canister and others of a cluster of several small canisters. One three-canister 227-kg model has a combined fuel content of 98 kg. The steel case of the canisters of these bombs is thin. The bombs are filled with ethylene oxide or other readily volatile and combustible liquid (such as propane, butane or other more exotic compounds or mixtures), under pressure in some of the models.

[3] For a detailed examination of the chemical and physiological properties of "CS" (*o*-chlorobenzalmalononitrile) and an *entrée* to the pertinent literature, *see* Jones (1972). *See also* the reviews by SIPRI (1971–1975: Vol. I:185–209; Vol. II:45–46).

Neilands (1972a) provides an extensive review of the use of CS by the USA in Indochina, and there are a number of additional sources for such information (Blumenfeld & Meselson, 1971; Blumenthal, 1969; Hersh, 1968:167–86; SIPRI, 1971–1975:Vol. I:185–209; Rose & Rose, 1972).

It can be noted here in passing that CS is often referred to as a gas although this is not the case. It is, in fact, a solid which is dispersed as an ultra-fine powder (aerosol). When CS is manufactured in finely pulverized (micronized) form it is referred to as CS1, and when the latter in turn is made water-repellent (and thus less rapidly degradable under field conditions) it is known as CS2.

[4] The best sources of information on military weather manipulation by the USA in Indochina are a series of US Senate documents (Pell, 1972a; 1973a; 1974). A number of interesting news accounts have also appeared (J. Anderson, 1971; Cohn, 1972; Greenberg, 1972b; Hersh, 1972; Shapley, 1972b; 1974). US military involvement in weather modification in general (exclusive of Indochina) has been described several times (Kotsch, 1968; Pell, 1972a; Studer, 1968–1969).

The literature dealing with intentional weather modification for civil purposes has been compiled *in extenso* (Grimes, 1972; Taborsky & Thuronyi, 1960; 1962; Thuronyi, 1963; 1964). There are also a series of articles that explore the ecological ramifications of such weather modifications, among them those by Cooper & Jolly (1969), Livingstone *et al.* (1966), Sargent (1967), Waggoner (1966), and Whittaker (1967).

[5] The fragmentary information available on the military use of fire during the Second Indochina War has been gathered and analyzed by SIPRI (1975:49–63). *See also* Neilands (1970b; 1972b; 1973), Takman (1967), and Lohs (1973b). The US Department of Defense has released virtually no information on its incendiary operations in Indochina nor on the levels of incendiary weapon expenditures there.

One can turn to Björnerstedt *et al.* (1973) for an excellent primer on incendiary weapons in general, and to SIPRI (1975) for a definitive monograph on the subject. These publications emphasize the anti-personnel and anti-matériel aspects of incendiary weapons and the social ramifications of their employment.

Brown & Davis (1973) provide a comprehensive coverage of forest fires from the standpoint of forestry (with major emphasis on conditions as they exist in the USA). The ecological effects of forest and other wild fires have been reviewed by Ahlgren & Ahlgren (1960), Broido (1963), Cooper (1961), Kozlowski & Ahlgren (1974), Lutz (1956), Mobley (1974), and others. A number of relevant bibliographies also exist (Cushwa, 1968; Hare, 1961).

Chapter 6. The ecology of disturbance

Where indicated thus,[1] the reader is referred to the notes on page 80.

I. *Introduction*

Various of the approaches to war which have been dwelt upon in previous chapters have had in common a devastating impact over wide areas. Indeed, such areas of devastation in some instances extended over hundreds, if not thousands, of contiguous hectares and thus represent a significant fraction of the recipient region. It therefore becomes necessary to explore the ecological consequences of such massive disturbance – consequences that, in fact, spilled over both the spatial and temporal boundaries of the attack.

The subject matter of this chapter falls more or less clearly into one or the other of two general categories: (*a*) the ecological or environmental consequences of disturbance (sections II – VII), and (*b*) the rate and ultimate extent of recovery of such disturbance (section VIII). Certain aspects of these consequences have, of course, been described already in earlier chapters in conjunction with the weapons themselves, especially those peculiar to a particular form of attack. Here it will be useful to discuss the shared and more general aspects of the subject.

Habitat disturbance is a dominant fact of life on earth. Ecosystems have been subjected to more or less drastic disturbances throughout the long history of their existence. Some disturbances, such as changes in climate, have been slow and subtle. Other disturbances, such as fires, tornadoes, earthquakes or tidal waves, have been sudden and spectacular. During modern history, man has been contributing to these disturbances at an exponentially increasing tempo, and there is a growing literature of concern.[1]

Following massive disturbance of a large area – whether it be caused by carpet bombing (chapter 2), herbicidal attack (chapter 3), Rome ploughing (chapter 4), or some combination of these or other techniques – one must expect changes of some magnitude and duration in the ecosystems involved. In considering these changes, one can distinguish among those occurring to the living component (or biomass) of an ecosystem, also often referred to as the "community" (sections IV and V), to its non-living component (or geomass) (section II), and to the nutrient (or biogeochemical) cycles that serve to connect these two major components (section III).

Within the biotic community itself, the further down one goes along the food chain, the more widespread and important the ultimate ramifications of a disturbance are apt to be. Thus, even if the direct assault were restricted to the autotrophic component of an ecosystem (that is, to the green plants), there would also be a profound secondary effect upon the heterotrophic component, both the herbivores and carnivores. However, having made this point, it becomes important to go on to stress that the complexity of reciprocal relationships and feedback mechanisms among all of the components of an ecosystem suggests that an assault on any single subcomponent (even if somehow restricted to the upper reaches of a food chain), could, in fact, turn out to be serious. In addition, the more disturbed or stressed an ecosystem already is as a result of past events – intentional or unintentional, military or civil – the more prone it will be to subsequent upset.

How does one go about determining the results and ramifications of any particular habitat disturbance, to say nothing of evaluating the significance of such changes? There appears to be no completely satisfactory answer to this question. However, one useful way of approaching the subject is to consider the environmental role or effect of a particular component of an ecosystem as this might be elucidated by its absence. In the past at least, such a difference between presence and absence has been a matter of some interest with respect to forest cover. It is the approach leaned upon in the present considerations of hydrology (section VI) and weather (section VII). These discussions are in turn followed by considerations of the tendency for a disturbed ecosystem to return to its former composition ("ecological succession"), in particular how rapid and complete this process is (section VIII).

Finally, it seems important to stress again that the consequences of an assault on the environment are by no means confined to either the locale or time of such attack. As will be elaborated upon below, when one area is attacked, another one far removed can be disturbed secondarily through flooding, silting, nutrient enrichment and other ways.

II. *The soil*

It is easy to visualize the soil as the Achilles heel of the terrestrial ecosystem.[2] Far more than an inert substratum, the fragile soil system, in fact, provides the crucial link between an ecosystem's living and non-living components. When all goes well, the modest continuing losses of soil material are compensated for by a constant process of soil genesis.

The continuing loss to an ecosystem of soil via running water or wind is known as erosion. Erosion is a natural process of universal occurrence, one which has operated throughout the history of the earth, sculpting its surface through the eons. The concern here, of course, is with accelerated erosion: soil removal at a rate in significant excess of soil genesis. Accelerated erosion has been and continues to be a serious problem in many parts of the world. Its economic importance is reflected in the vastness of the literature on the subject.[2] Indeed, V. Carter & Dale (1974) have linked the very birth and death of civilizations to the regional soil status. It must be noted, moreover, that problems of erosion are linked with those of flooding (section VI) and with problems of nutrient losses in solution (or nutrient dumping) (section III).

Various factors influence the type and rate of erosion, among them several aspects of the climate, a number of soil characteristics, the slope and other terrain features, and – most importantly – the status of the vegetative cover and its associated litter. A vegetative cover protects against erosion in a number of major ways. The vegetation and the litter beneath it (the so-called L or A_0 horizon) together buffer the soil from the direct impact of rain. The plants also ameliorate the force (and thereby the carrying capacity) of the wind. The interlocking root systems of the plants are important in holding the soil in place. Both the surface litter and the decomposing organic matter (or humus) have high water-holding capacities. Moreover, the upper horizons of a soil beneath vegetation develop a high porosity as roots die and decompose and as worms and other herbivorous subterranean creatures tunnel their way about. This provides for subsoil infiltration (or cutoff) rather than erosive surface flow or runoff. For example, Lowdermilk (1930) found that erosional soil loss can readily increase by a factor of more than a thousand following the vegetational and litter destruction that results from a forest fire (*see also* H. Anderson *et al.*, 1966; Arend, 1941).

The primary damage resulting from accelerated erosion is both obvious and serious: the removal of all or part of the litter and topsoil. This particulate loss in turn results in the loss of the incorporated organic matter and nutrients (section III). It further leads to soil compaction, a deterioration of soil structure, a reduction in water-holding capacity, a decrease in the soil fauna and in related deleterious effects. What all of this adds up to is decreased site productivity of long duration.

Another major problem with accelerated erosion is the off-site damage. Some of this occurs during transport (whether it be through wind or water), although most is at the point of deposition. In the case of water erosion, siltation can bring about severe habitat debilitation in both streams and lakes by disrupting fish spawning grounds and in other ways (D. Chapman, 1962; Ellis, 1936). Silted-in stream channels in turn can lead to increased flood damage. Both water- and wind-carried materials are also known to become deposited on terrestrial habitats, thereby sometimes killing the extant vegetation and causing other damage.

A number of routine military activities lead to seriously accelerated soil erosion. For example, substantial damage from soil erosion by water occurred on a number of World War II battlefields in the Soviet Union, some stemming from battle scars *per se* and others from trenches and tank traps (Sobolev, 1945; 1947; Sus, 1944). When the vegetational cover is disrupted in the hilly (sloping) terrain of Indochina, the soil is particularly subject to erosion during the torrential downpours characteristic of that region. Substantial erosional damage resulted from the Rome ploughing, the bombing and shelling, and the use of herbicides. Particularly severe erosional damage was observed to result from the spraying of the herbicides in the mangrove regions, a matter described elsewhere (page 38). Wind erosion can also result from military activities. For example, Sterba (1970) has observed dust storms in South Viet-Nam stemming from the US military activities there.

Certain soil types in the very wet and hot regions of the world such as northern South America, central Africa and southeast Asia have the curious tendency to become hard upon prolonged exposure to the elements. This process of hardening – referred to as "laterization" or "induration" – appears to be nearly irreversible once it has run its course. With the massive habitat disturbances inflicted upon the land of Indochina by the war, the spectre of laterization arose early on. Fortunately, however, the occurrence in Indochina of soils capable of at least rapidly indurating is relatively uncommon and scattered. This, coupled with the usually quite rapid recolonization of soil laid bare (*see* page 76), seems to have precluded any widespread formation of laterite.[3]

In conclusion, the Second Indochina War can be seen to have resulted in extensive damage to the soils of the region, and thereby to the ecosystems involved. But to quantify this damage or otherwise suggest its level of importance appears not to be possible.

III. *Nutrient cycling*

The biogeochemical cycling of mineral nutrients is one of nature's fundamental phenomena and is considered to be one of the cornerstones of the ecosystem concept.[4] The bulk of any particular nutrient presumably cycles repeatedly between the biomass and geomass of an ecosystem via a more or less complex set of pathways. Even under ideal conditions, however, a small – though significant – fraction continues to leak out of the system, in turn to be replenished by a comparable input. The rates of loss and replenishment as well as the major source of replenishment (the so-called pool or reservoir) differ from nutrient to nutrient. Moreover, each of the nutrient cycles differs in its ability to withstand or recover from outside disturbance. For example, the nitrogen cycle is considered to be relatively more robust than the phosphorus cycle.

One of the subtle, though serious, consequences of vegetational disturbance in an ecosystem is the disruption of its nutrient cycles. Such disruption manifests itself in the rapid loss to the site of a large quantity of mineral nutrients. Some of this loss can be attributed to accelerated soil erosion (that is, to loss of particulate matter) (section II) and some to increased water movement (section VI). There occurs, however, a substantial additional loss of nutrients in dissolved form even when the runoff (surface flow) and cutoff (deep percolation) remain at more or less normal levels. This phenomenon can be termed "nutrient dumping" (Westing, 1971b).

The severity of nutrient dumping hinges not only on the character and degree of disturbance, but also on such site factors as temperature, precipitation, soil type and slope. As will be seen, nutrient dumping is a potentially serious form of terrestrial ecosystem debilitation anywhere. However, there are indications that it may be substantially more of a problem in the tropics than in temperate zones.

The phenomenon of nutrient dumping has been studied in greatest detail as an outgrowth of a major nutrient-budget study in a mature dicotyledonous forest in New Hampshire, USA (Bormann *et al.*, 1968; Likens *et al.*, 1970; Bormann *et al.*, 1974). The investigators were able to quantify both the particulate (erosional) and solute losses resulting from the felling and subsequent poisoning of all vegetation on a 16-ha watershed. It should be noted that they left all of the cut and killed vegetation in place and that care was taken not to disturb the soil surface and its associated litter.

The nutrients that flushed out of the system originated in the litter from which they had been released by accelerated microorganismal decomposition. Bacteria produced large quantities of soluble inorganic nitrogen in the form of nitrate ions (NO_3^-), a chemical species that the soil is essentially incapable of holding. The microorganisms concomitantly generated large quantities of hydrogen ions (H^+) which in turn served to displace various nutrient cations (such as K^+, Ca^{++}, Mg^{++}) which had been adsorbed (bound) primarily onto the colloidal organic and clay fractions of the soil. Large quantities of nutrients were in this manner lost to the disrupted ecosystem within a short period of time. It can be noted here also that McColl & Grigal (1975) recently reported the greatly accelerated loss of phosphorus from a forest ecosystem following its disruption by fire.

In a critique of the New Hampshire study, Reinhart (1973) has pointed out that following normal logging operations the soil could still contain substantial levels of more or less readily available nutrients. Reinhart's hesitation to relate the results of the New Hampshire study to the ecological impact to be expected from routine forestry operations is, of course, not applicable in the present context inasmuch as the severity of military disruptions of an ecosystem might even exceed those that were experimentally created in the New Hampshire study.

As vegetational regrowth begins on a disrupted site, nutrient dumping is arrested (as are accelerated water loss and erosion). This sets in motion a more or less rapid return to normalcy with respect to nutrient cycling (Bormann *et al.*, 1969; Bormann *et al.*, 1974; Marks & Bormann, 1972). Lost nutrients are slowly resupplied to the site in a variety of ways, some more important than others for the various nutrients. Thus, some ions are derived largely from the slow dissolution of the rock or parent material dispersed in and beneath the soil, a process referred to as weathering. The reservoir for nitrogen, on the other hand, is the atmosphere, from which the rather inert gaseous nitrogen (N_2) is oxidized or "fixed" to a form available to the higher plants. Such fixation is performed by a limited array of free-living and symbiotic microorganisms, some found in the soil and others in intimate associations with roots or leaves. Another seemingly important, though generally unappreciated, source of steady resupply of various nutrients is the atmospheric fallout, both wet and dry, that is deposited on the site.[5] And of course, in some special localities, such as broad mature stream valleys, the site can be replenished seasonally with nutrient-rich deposits of silt from surface flow.

Nutrient dumping may be a particular threat in the wet tropics. In a temperate forest ecosystem less than half of the total nutrients are to be found in the biomass at any one time. In a tropical forest ecosystem, on the other hand, more than three-quarters (perhaps substantially more) are found in the biomass (Golley *et al.*, 1969; Kira & Shidei, 1967; Nye & Greenland, 1960; Richards, 1952:219–21; Yoda & Kira, 1969). In the tropics, moreover, the litter layer is thin and the soil is notorious for its poor nutrient-holding capacity. Indeed, it has even been suggested that normal nutrient cycling in a tropical forest may bypass the non-living component of the soil system almost entirely (Stark, 1971; Went & Stark, 1968).

Once a tropical forest stand is killed, the resulting nutrient-rich organic debris decomposes rapidly, far more so than its temperate-zone counterpart (Jenny *et al.*, 1949; Madge, 1965). Indeed, in the tropics the rate may be as rapid as 1.3 per cent per day (Nye, 1961). With a paucity of adsorptive (binding) sites in the tropical soil complex to begin with (and these perhaps newly occupied in part by H^+ ions) the nutrients being released by organic-matter decomposition are rapidly leached out of the soil, to be lost to the disrupted ecosystem in the runoff and cutoff (Cunningham, 1963; Nye & Greenland, 1964). The greatly impoverished site will then be able to support only a limited array of pioneer vegetation (section VIII). Substantial nutrient recovery via natural processes may then take a decade or longer in the tropics (Nye & Greenland, 1960).

Thus it is seen that nutrient dumping can be a major problem associated with ecosystem disturbance in both temperate and tropical forests, especially so in the latter instance. It becomes evident that nutrient dumping must become a particular threat in conjunction with massive military disruptions of the vegetational cover. Indeed, this has been found to be the case in the only directly relevant study to date. Zinke (1974:16–25) was able to compare the soil from a dense upland forest in South Viet-Nam which had been herbicidally attacked seven years previously with a comparable control soil. He found the war-disrupted site to possess severely diminished levels of nitrogen and phosphorus and also to be significantly more acid in reaction.

It must further be pointed out that nutrient dumping may well lead to additional problems beyond the locus of attack. The nutrients flushed out of the disrupted ecosystem might then contribute to an undesirably high nutrient enrichment of another ecosystem elsewhere. The latter is likely to be an aquatic ecosystem, the excessive enrichment then usually being referred to as accelerated eutrophication (Likens & Bormann, 1974; McColl & Grigal, 1975).

Thus it can be concluded that the multifariously war-disrupted ecosystems of Indochina suffered severe, though largely hidden, debilitation via nutrient dumping. However, the overall level of this form of damage seems even more difficult to quantify than that attributable to the soil erosion examined elsewhere (section II).

IV. *The vegetation*

Vegetation on the battlefield is subject to severe abuse. This became particularly so as a result of the several indiscriminate methods of wide-area attack which were developed in Indochina. Indeed, the chemical anti-plant agents and Rome ploughs employed so extensively by the United States during the Second Indochina War were designed specifically to accomplish massive vegetational destruction. The same can be said for the concussion bombs used, though on a far lesser scale, and might have been said for the forest fires attempted, had it been possible to develop satisfactory techniques.

The vegetation can be divided conveniently into trees on the one hand, and all that remains on the other. Whereas at least a rough estimate of damage can be made for the trees, it is not feasible to attempt this for the remaining flora. This is, of course, an unfortunate circumstance since an estimate restricted to tree damage will ignore most of the extant vegetation – including, *inter alia*, the herbaceous plant species, the lower plant species, the aquatic plant species and the agricultural and horticultural plant species. On the other hand, for many of the sites of interest in the present context, a tree analysis will account for almost all of the vegetational biomass, indeed, for almost all of the entire biomass, plant and animal.

The vegetation in a theatre of war can suffer either direct battle damage (that is, resulting from so-called hostile action) or various forms of indirect damage. For example, the Japanese are said to have exploited ruthlessly the forests of Indochina during their occupation of that region during World War II (McKinley, 1957:1, 45). However, in the analysis that follows only direct damage to the forest resource is considered. Moreover, the analysis is restricted to the forest damage which occurred in South Viet-Nam, the most heavily war-abused region of Indochina (table 1.4).

About 10.4 x 10^6 ha or 60 per cent of South Viet-Nam is forested (table 1.2). The 5.9 x 10^6 ha of this area considered to be commercial forest is sufficiently large and productive to supply the country with perhaps twice its annual timber needs of roughly 1.5 x 10^6 m^3 (Westing, 1971c).

These forests have been abused in a multitude of ways. Past chapters have examined the overall ecological impact on all of Indochina of bombs and shells, chemical anti-plant agents, Rome ploughs and miscellaneous weapons and techniques. Presented below are extractions and extrapolations from those previous presentations only insofar as they apply to the trees of South Viet-Nam.

The damage to South Viet-Nam's forests from bombs and shells is most conveniently presented in two stages, complete obliteration and severe damage. The first category consists of that forest land which was converted to craters by the high-explosive munitions. Using the more or less arbitrary assumption that in South Viet-Nam a forested area was twice as likely to be a bomb or shell target than a non-forested one, it can be calculated from the data presented in table 2.5 (reduced according to its note [e]), that such crater-obliterated forest areas add up to about 104 x 10^3 ha (or, if one wishes to consider only the commercial forest, to about 59 x 10^3 ha). Next there is the forest land that was subjected to flying metal fragments ("shrapnel"). If one uses the zone subjected to such abuse at an intensity sufficient to be lethal to 50 per cent or more of exposed personnel (*see* note [3], page 22), then the area in question amounts to about 4.9 x 10^6 ha (or for the commercial forest, to about 2.8 x 10^6 ha). This last defined area is one in which many of the trees are injured by "shrapnel", an event that in turn leads to fungal entry and decay, inevitably followed by a significant proportion of tree mortality.

The damage to South Viet-Nam's forests from chemical anti-plant agents is also most conveniently presented in two stages, virtually complete obliteration and partial damage. The first category consists of that forest land which was sprayed four or more times if an upland area and once if a mangrove area. It can be calculated from the data presented in tables 3.4 and 3.6 that this category of virtual obliteration covers some 51 x 10^3 ha of upland forest (or of commercial upland forest, about 29 x 10^3 ha) plus an additional 151 x 10^3 ha or so of mangrove forest. The second category consists of upland forest land sprayed one to three times. This area of partial damage covers some 1 266 x 10^3 ha (or of commercial upland forest, about 718 x 10^3 ha). It is estimated that the first category experienced between about 85 and 100 per cent tree mortality whereas the second category experienced between about 10 and 50 per cent (*see* page 31).

The damage to South Viet-Nam's forests from Rome ploughs need be represented in but one category, that of essentially complete obliteration. The category amounts to about 325 x 10^3 ha (or in terms of commercial forest, to about 184 x 10^3 ha) (*see* page 47).

The damage to South Viet-Nam's forests from miscellaneous hostile actions is being ignored in the present estimations (as is that from various military actions not directly battle-related) although these diverse sources might well have contributed several thousands of hectares to the total.

Combining the several separate damage estimates made above for the forests of South Viet-Nam is made difficult by an inability to determine with any degree of accuracy the extent to which any one area was subjected to more than one of the several categories of insult. In the following summations the extent of damage has been reduced by 10 per cent to account for such overlap. Thus, complete or essentially complete devastation comes to an estimated 568 x 10^3 ha, representing 5 per cent of the total forest lands of South Viet-Nam. Considering only the commercial upland forest, this value becomes 245 x 10^3 ha, or 4 per cent of that category. The partially (severely) damaged forest lands are estimated to amount to an additional 5.6 x 10^6 ha (or for the commercial forest, 3.2 x 10^6 ha), or 54 per cent of these categories.

It is further possible to make a crude approximation of the merchantable timber losses in South Viet-Nam that resulted from the three forms of attack discussed above. If one assumes that the standing merchantable crop on the commercial forest averages 90 m^3/ha (Rollet, 1962:4; Tân, 1971) and if one further assumes that the portions designated above as obliterated (or essentially so) in fact sustained a 95 per cent loss and the portions designated as partially (severely) damaged sustained only a 20 per cent loss, then the war damage examined here amounted to a merchantable timber loss of about 74 x 10^6 m^3, that is, to about 14 per cent of the nation's standing merchantable timber crop. Losses of fuel wood and other secondary products are, of course, not included in this figure. Indeed, the host of ancillary and intangible war losses to the forest resource and other vegetation are not considered here, largely for lack of relevant data.

Thus, although the estimate of vegetational damage made here has been restricted to the trees of South Viet-Nam, it should be safe to assume that various components of the flora throughout Indochina sustained greater or lesser amounts of damage as well. The botanical forces set in motion by such disruption are examined elsewhere (section VIII).

V. *The fauna*

Warfare can have a variety of effects on animals ranging from intentional to unintentional, from direct to indirect, and from favourable to unfavourable. Some wildlife is bound to become a direct casualty of attack when pattern bombing or blast munitions are resorted to. However, some of the larger wildlife is spared from such direct injury by its ability to flee from areas of military commotion. For example, in the relatively peaceful eastern Cambodia of 1969 the populations of muntjacs (*Muntiacus muntjak*, Cervidae) and other species of deer, of wild cattle such as gaurs (*Bos gaurus*, Bovidae), bantengs (*B. banteng*), and some koupreys (*B. sauveli*), of elephants (*Elephas maximus*, Elephantidae), of a number of monkey species (Cercopithecidae), and of wild pigs (*Sus* spp., Suidae) had all recently increased to relatively high levels, refugees from the fighting in neighbouring South Viet-Nam (Westing, 1972a:197). For many animals, however, such escape is not possible since the undisturbed areas of potential refuge are already likely to be populated to their maximum carrying capacity.

Clearly the most serious impact of military hostilities on the fauna is an indirect one, via their impact on the faunal habitat. Thus any of the various operations that decimate the vegetative cover concomitantly decimate the sources of food and shelter of the associated animal life. This obligate relationship between an animal and its *milieu* is a fundamental tenet of wildlife biology (A. Leopold, 1933; Allen, 1962; Wing, 1951). It becomes even more evident in tropical forests because an especially high proportion of the tropical fauna is restricted to the treetops (Allee & Schmidt, 1951:511–14, Richards, 1970; Ripley *et al.*, 1964). As Audy (1948) so vividly put it, forest destruction in the tropics converts a rich three-dimensional habitat to a depauperate two-dimensional one.

It should thus come as no surprise that the destruction of wildlife habitat has been one of the sad concomitants of battle damage in Indochina. This was especially obvious in the mangrove estuaries that were laid waste during the war. As is described elsewhere (page 38), the normally rich avian fauna that populates the mangroves was decimated. Moreover, population levels of the aquatic fauna, both vertebrate and invertebrate, that depend upon this habitat for all or part of their life cycle were also found to have been significantly reduced.

Ngan (1968) has listed five animals as being on the verge of being extirpated in South Viet-Nam: the kouprey (*Bos sauveli*, Bovidae), a tapir (*Tapirus indicus*, Tapiridae), a bear (*Ursus tibetanus*, Ursidae), a gibbon (*Hylobates pileatus*, Pongidae), and a pheasant (*Lophura imperialis*, Phasianidae). It is, in fact, the fear of some Indochinese zoologists that the Second Indochina War may have provided the *coup de grâce* to the already highly endangered kouprey (*see also* Curry-Lindahl, 1972: 149–50). Two additional animals should also be mentioned here whose well-being if not existence has become threatened by this war. According to local icthyologists, the large fresh-water tarpon *Megalops cyprinoides* (Megalopidae) seems to have become at least locally extirpated in the Mekong delta region of South Viet-Nam. There are also indications that the clam *Polymesoda coaxans* (Corbiculiidae) found in the mangrove estuaries of South Viet-Nam may have been placed in jeopardy by the herbicidal attacks against that habitat. The extirpation of an animal species sometimes results in spectacular perturbations within the residual community, whether it be terrestrial (Paine, 1966; Rasmussen, 1941) or aquatic (Estes & Palmisano, 1974).

The vegetational community that colonizes a severely disturbed site will support an animal community far different from the original one. It will consist of fewer species, many of them new to the area (and in human terms often considered to be undesirable). For example, it appears that rodent population levels increased dramatically in South Viet-Nam in some of the herbicide-decimated areas (Desowitz *et al.*, 1974; Lang *et al.*, 1974:VII:13, 21). Similarly in Malaysia, it has been observed that the destruction of indigenous tropical rain forest resulted in the virtual elimination of the native mammalian fauna, with introduced rats (*Rattus*, Muridae) being the most prominent residual species (Harrison, 1968).

Two additional members of the Indochinese fauna can be mentioned. In the first instance, there are persuasive indications that the widespread military destruction of the mangrove habitat in South Viet-Nam has led to increased numbers of the malaria vector mosquito *Anopheles* (Culicidae) (Desowitz *et al.*, 1974). This increase may in part also be attributable to the many millions of water-filled bomb craters which provide breeding grounds with reduced predator populations for this insect (Westing & Pfeiffer, 1972). Secondly, there exists a macabre report that tigers (*Panthera tigris*, Felidae) in South Viet-Nam were thriving on the widely available dead and wounded soldiers (Orians & Pfeiffer, 1970:533).

Thus it is clear that the fauna of Indochina underwent severe disturbance as a result of the war, most of it probably stemming from habitat disruption and much of it presumably undesirable. How one evaluates the significance of such perturbations is hard to say.

VI. *The hydrology*

The hydrologically protective and stabilizing influence of vegetation is well known: a watershed supporting a healthy and well-stocked vegetative cover produces fewer and less severe floods than one on which the vegetation has been disrupted.[6] The present section amplifies upon this statement and discusses its implications as these relate to the vegetative disruptions brought about by military actions. Related to severity of flooding are the severity of erosion and of nutrient losses in solution, that is, nutrient dumping, but these factors are dealt with elsewhere. Also covered elsewhere is intentional flooding caused by the military destruction of dams and so forth (*see* pages 56–58).

When the vegetation on an area and its associated litter are severely disrupted, the flood danger becomes markedly increased. Thus, it has been demonstrated a number of times that when the vegetative cover on a watershed has been experimentally destroyed, this results in greatly increased stream flow out of the watershed, both at peak times and overall (Douglass & Swank, 1972; Hornbeck *et al.*, 1970; Likens *et al.*, 1970; Patric, 1973). Various natural disasters have been shown to produce similar results, including the vegetational disruption wrought by hurricanes (Patric, 1974), fire (H. Anderson *et al.*, 1966), or insect attack (Bethlahmy, 1974). There are a number of reasons for this. For one thing, overland flow (surface runoff) is stimulated, a phenomenon associated primarily with site disruption through fire or mechanical means (Arend, 1941; Lowdermilk, 1930). Secondly, when the litter is lost, its water-storage capacity is lost with it.

A most important reason for increased flood potential following vegetative destruction is the resulting reduction in water loss from the site via transpirational evaporation (flyoff). This is reflected in wetter soils and a substantial rise in the level of the water table following forest clear-cutting (Troendle, 1970; Trousdell & Hoover, 1955; Wilde *et al.*, 1953). With the storage space of the soil thus occupied, subsequent rainfall may be forced to vacate rapidly the affected area via surface or subsurface flow.

The period of greatest concern is, of course, when the soil is exposed, that is, before at least some vegetative cover becomes re-established on the site. Interestingly enough, it does not seem to make much difference with respect to flood amelioration whether the replacement vegetation is arboreous or herbaceous. Thus, Hibbert (1969) could find little difference between stream flow out of an experimental watershed following the replacement of the existing dicotyledonous forest cover by a stand of healthy, well-fertilized grass (Gramineae). This seems explicable on the basis that both types of vegetation transpire roughly comparable quantities of water. Lull & Reinhart (1972) however, suggest that forest cover might provide somewhat better flood protection than grass cover, primarily because there develops a greater capacity for water storage in the soil. This is presumably brought about by a greater rate of evaporation resulting from a combination of lower albedo, deeper root systems and perhaps other factors (Troendle, 1970). Moreover, the results of some investigations in Hawaii suggest that at least there forest soils possess a somewhat higher water-infiltration rate (H. Wood, 1971) and a somewhat greater water-holding capacity (Yamamoto & Duffy, 1963) than do grass and other non-forest soils.

No specific observations appear to be available regarding the aggravation of flood conditions as an incidental result of military disruption of the vegetative cover. It becomes abundantly evident from the foregoing discussion, however, that one can expect this to occur when such techniques as Rome ploughing, pattern bombing, herbicide spraying or the successful employment of forest fires are used, either singly or in combination. Such increased flooding problems will persist downstream from the disrupted watersheds during the months or years that it takes for the re-establishment of an adequate plant community.

Finally it should be noted that intense bombing of an area might modify local drainage patterns and that bomb craters might exert an influence on the water table.

VII. *The weather*

This section covers the inadvertent or ancillary weather changes that might result from military activities. The intentional manipulation of the weather for hostile military purposes is covered elsewhere (pages 55–56).

Despite the erratic nature of meteorological fluctuations, it is now known that man has inadvertently changed regional weather patterns through various of his activities (Bryson, 1974; P. Hobbs *et al.*, 1974; Landsberg, 1970; Machta & Telegadas, 1974). For example, highly urbanized and industrialized areas exhibit altered local climates (Changnon, 1968; Lowry, 1967) and widespread irrigation is reported to result in regional increases in summer rainfall (Stidd, 1968). However, clearly the most pronounced changes that can be attributed to man are in the climate near the ground, or microclimate, of disturbed areas (Geiger, 1965; Glesinger, 1962:III; Hughes, 1949–1950:I; Kittredge, 1948; Toumey & Korstian, 1947:XI).

It has long been a part of military folklore that major battles bring about subsequent rainstorms, provoked somehow by heavy artillery fire. Powers (1890) attempted to marshal anecdotal evidence in favour of this notion from the US Civil War and several others, but was unconvincing. In the present context, the question arises whether military habitat disruption over large areas, for example, through the herbicide spraying or Rome ploughing that was carried out in South Viet-Nam, has had any appreciable meteorological influence.

Bryson & Baerreis (1967) provide perhaps the most dismal possibility. They suggest that the Rajputana desert in northwestern India might be the result of climatic changes stemming from a series of events that had originally been set in motion by large-scale vegetational removal by the local inhabitants of the region. The case they present appears to have some merit and has, moreover, been supported by some interesting recent observations in the Middle East (Otterman, 1974). However, Bryson & Baerreis themselves admit that the matter requires further study. At the other extreme, one can point to the preliminary evapotranspiration calculations of Tschirley (1969) which have demonstrated to his satisfaction at least that the removal of a block of tropical forest 10^6 ha in size would have no significant effect on the regional rainfall. By way of contrast, Newell (1971) has put forth the suggestion on the basis of his calculations that tropical deforestation could, indeed, have an appreciable influence on the climate.

Perhaps the most important actual study of relevance to the present discussion involves the Ducktown or Copper Basin region in southeastern Tennessee, USA. It is here that copper smelting fumes literally destroyed the forest vegetation on almost 10×10^3 ha (Seigworth, 1943). This environmental insult resulted in a totally bare inner area about 2 800 ha in size surrounded in turn by a grassy region of perhaps 6 900 ha. In a very detailed study, Hursh (1948) collected a variety of meteorological data in this region over a period of four years. This enabled him to compare the microclimates of three contiguous land areas – forested, grass-covered and absolutely bare – all of which had originally supported an essentially uniform forest stand.

As is obvious, Hursh recorded substantial differences among the sites in air and soil temperatures, wind velocity, evaporation and other factors of the microclimate. What came as somewhat of a surprise were the differences he recorded in local rainfall in the three contrasting sites. Statistical analysis demonstrated highly significant differences in rainfall among all three of the sites. During the four years of the study the average rainfall on the surrounding forest was 1 460 mm/year as compared with 1 340 mm/year on the grassy zone (a reduction of 8 per cent) and 1 280 mm/year on the bare zone (a reduction of 12 per cent). Hursh suggested that an explanation for these differences might be found in the differing albedos of the three sites. These differences in reflectivity would in turn account for differences above the sites in air temperature, density, relative humidity and turbulence – together presumably sufficient to bring about the differences in observed rainfall.

Thus it is evident that large-scale military habitat disruption will profoundly alter the microclimate of the disturbed area. Particularly affected would be the extremes of insolation, temperature, wind and moisture on and near the ground. These changes in turn are able to contribute to the remarkable differences between the biotic community that will colonize a disturbed area and the one that the area supported prior to the disturbance (*see* below). On the other hand, it is somewhat less clear whether vegetational disruption of the magnitude that occurred during the Second Indochina War has had an appreciable influence on the regional weather there (*see also* Emmons, 1971–1972; Kutzleb, 1971–1972; McDonald, 1962).

VIII. *Recovery*

Destruction or severe disturbance of an ecosystem sets in motion a more or less orderly, and thus predictable, series of events. This progression – known as ecological succession – involves as a first step the establishment on the site of a new biotic community, one which is quite different from the destroyed one. Over a period of years or decades this colonizing community, or pioneer stage, gives way to a series of more or less distinct replacement communities, known as intermediate stages. Eventually, if all goes according to theory, a community becomes established that perpetuates itself on the site, the so-called climax stage.[7]

The reason that the pioneer community is so different from the pre-disturbance community is that the local habitat has been more or less drastically altered in a number of important ways, as has been described in earlier sections. To recapitulate, the microclimate has become remarkably different, the site, *inter alia*, now exposed to the direct rays of the sun, now much windier and thus more highly evaporative, and now experiencing greater extremes and more rapid changes of temperature and moisture. A combination of erosion and nutrient dumping have served to diminish greatly the nutrient levels in the soil, a soil whose structure and water-holding capacity have also deteriorated. Moreover, wildlife food and cover – and thus also the wildlife itself – have been largely eliminated.

Beyond the necessity for their local availability, the colonizing pioneer plants must be photophilic (that is, the foresters' intolerant) and able to cope with the whole diversity of newly established site rigours. Once established, these hardy pioneers – by the simple virtue of their presence – serve to change the local site conditions. In doing so, they make the local habitat relatively more suitable to a new mix of plant species, ones that are somewhat more tolerant (*sensu* forestry). The pioneer species bring about the site changes by providing partial shade to the ground surface, by reducing the ground-level wind velocity, by beginning to hold the soil in place with their roots, by providing the land with the beginnings of a protective litter layer, and so forth. They also provide new niches (that is food and shelter) to a limited selection of animal life, both above- and below-ground.

As mentioned above, this drama of succession unfolds in a relatively predictable fashion, eventually culminating in a climax stage. The nature of this steady-state community differs, of course, from region to region, depending in large measure upon the regional patterns of rainfall and temperature. The climax biota maintains a dynamic equilibrium with the soil (which it has helped to shape) and the climate. Presumably this stage approximates the so-called old-growth or primaeval situation.

Ecological succession is, of course, a worldwide phenomenon, occurring in all climates in both terrestrial and aquatic habitats. The entire biotic community – indeed, the entire ecosystem – undergoes successional changes, with continuing interplay occurring between plant, animal and non-living components. It is, however, the plant component of the ecosystem that appears to dominate the events and seemingly relegates the other components to subordinate roles. Thus, animal succession for the most part appears to be dependent directly or indirectly on the vegetational succession.[7] The soil, with its originally diminished nutrient levels, reduced structure, and damaged profile, undergoes a concomitant recovery, also roughly in step with the vegetative succession (Lutz & Chandler, 1946:XI).

The length of time required to attain climax is, of course, highly dependent upon a number of factors, most important of which is the severity of disturbance which had originally set the changes in motion. Moreover, the regional climate not only controls the vegetational character of the climax community (various types of forest, prairie and so on), but thereby also the time span between pioneer and climax. Temperate-zone successional series culminating in some association or another dominated by grasses usually take several decades to run their course, whereas those culminating in tree communities can take several centuries. Moreover, it must be recognized that when one deals with time spans of the latter magnitude, successional development is likely to be complicated by and confounded with changes in climate and physiography or geomorphology. Furthermore, depending upon its location, the ecosystem is more or less frequently subjected to new disturbances as the years go by, both natural and anthropogenic. Indeed, such disturbances as fires, hurricanes, floods, insect irruptions or earthquakes have always been factors with which one ecosystem or another has had to cope with greater or lesser frequency.

One of the important trends in the successional development of an ecosystem from pioneer to climax stages is the continuing increase in the variety of the plants and animals, both large and small. And as species diversity increases, the complexity of the food chains and other species interactions also increases. Such changes make for ever-increasing ecosystem stability.

Both erosion and nutrient dumping are arrested early on in succession, and as time goes by, the various nutrient or biogeochemical cycles become ever more closed. Both the biomass and its detritus (the soil litter and other organic matter and their decomposition products) keep increasing, and various forces serve to enrich the nutrient capital of the system. As discussed previously, these include dissolution of the rock material dispersed in and beneath the soil, and also transport to the site via wind and water.

When an ecosystem is disrupted by military actions, successional recovery of such an ecological setback is, of course, set in motion. Two levels of site recovery must be distinguished here. On the one hand there is the establishment of a pioneer stand, and on the other the eventual return to the pre-disturbance situation – which, of course, may or may not have been pristine to begin with. The first of these – measurable in months or years – arrests the massive damage to the ecosystem which is occurring from wind and water erosion and from nutrient dumping. The second – measured in years, decades, or even centuries – involves more subtle and intangible aspects of recovery. It is important to note that discussions of recovery from military actions often fail to distinguish between these two levels of recovery. This has been particularly so in some of the uncritical literature about the Second Indochina War.

Where the vegetational cover is destroyed on the tropical upland sites of Indochina, the land soon becomes fully occupied with vegetation, often within months. The pioneer vegetation on such greatly disrupted sites is likely to be a relatively pure stand of grassy vegetation, sometimes woody but more often herbaceous. In South Viet-Nam and elsewhere in Indochina, the woody grasses (Gramineae) likely to colonize these disturbed areas are such frutescent (shrubby) bamboos as *Bambusa, Thyrsostachys* and *Oxytenanthera* (Drew, 1974a), whereas the most likely herbaceous grass is the notorious *Imperata cylindrica* (for information on *Imperata, see* Hubbard *et al.*, 1944). Severe disturbance, particularly in North Viet-Nam, may also invite invasion by such herbaceous non-grasses, or forbs, as *Eupatorium* (Compositae).

The tenure of these weedy species, known for their ability to survive on severely impoverished sites, may be quite lengthy. The stands of bamboo (known as "brakes") may readily dominate the site they have occupied for half a century or longer. Dicotyledonous forest tree species are presented with a possible opportunity to regain a foothold (that is, to take over as the next successional stage) only once every several or more decades, at the time of general flowering and culm (stem) death. Moreover, the tenure of the bamboo stage is prolonged by an occasional fire (Drew, 1974a). The occupancy of a site by these bamboo brakes can be so protracted that this pioneer stage of succession could at the same time be referred to as a pseudo-climax stage.

The *Imperata* grass may also become a semi-permanent feature of the landscape, particularly if its competition is knocked back by an occasional fire. Indeed, vast tracts of seemingly permanent tropical savanna are maintained in just this fashion (Budowski, 1956; Holmes, 1951; Wharton, 1966; 1968). If, on the other hand, fires do not occur on the site, experience in Sumatra suggests that forest can usually crowd out the *Imperata* within 10 to 15 years or so (Nye & Greenland, 1960:VII). But then again, according to Holmes (1951), if the tenure of this grass is extended for long periods it will then be able to maintain itself even with fires excluded.

The successional progress toward the tropical steady-state climax community has not been studied nearly as much as its temperate counterpart.[7] It may, in fact, be as slow in the tropics as it is in the temperate zones. For example, Meijer (1970) investigated a forest in northern Borneo (Sabah) some 40 years after logging and found it to be still remarkably different from the neighbouring uncut forest (*see also* Meijer, 1973). Taylor (1957) found the same to be the case for a site in northern New Guinea which had been drastically disturbed some 83 years prior to his observations. In the disturbed site there were only 110 different tree species present (of which 56 were climax species) as opposed to considerably more than 500 in the equivalent undisturbed forest nearby. Moreover, Chevalier (1948:110) found that a forest in north central Cambodia known to have been developing on cleared land which had been abandoned to nature some 500–600 years previously, although closely resembling nearby primaeval forest, still exhibited noticeable differences.

Turning now to the coastal mangrove habitat, a rather different picture emerges. When this association is destroyed – as was done extensively by herbicides in South Viet-Nam – the site does not readily become recolonized. A lack of adequate seed source, a destruction of available propagules by crabs (Brachyura), and other more obscure factors combine to convert such mangrove sites to a muddy wasteland into the indefinite future. Indeed, it has been calculated that substantial initial recovery can be expected to take more than a century (Lang *et al.*, 1974:IV:119; H. Odum *et al.*, 1974:289).

What can be done by man to hasten the process of ecological recovery on war-disrupted sites? To a large extent there is nothing esoteric about the actions that might be taken. Thus, craters can be filled in by hand or, as available and when the terrain will support it, with heavy equipment. Erosion can be prevented by a variety of standard mechanical and vegetational techniques of long standing.[2] Soil nutrients can be replenished by the establishment of leguminous cover crops and the application of fertilizers, both organic and inorganic. Old drainage patterns can be re-established, levees and dikes rebuilt and irrigation systems restored. And both orchard and forest trees can be replanted (C.W. Swanson, 1975; Westing, 1974b).

On the other hand, a variety of factors obstruct such efforts at reclamation. Chief among them, of course, are the extensiveness of the disruption and the paucity of human and material resources. Nevertheless, a few relevant remarks follow.

To begin with, it should be mentioned that there exists a growing body of literature dealing with the reclamation of spoil banks (slag heaps) and exposed subsoil left behind by strip (opencast) mining. It seems that some of the research results from this field might lend themselves as well to the reclamation of areas that have been subjected to heavy bombing or shelling.[8] Secondly, the tenure of the tenacious weedy pioneer vegetation that often colonizes massively disrupted sites in Indochina can be curtailed by a number of means. For example, Holmes (1951:52–54) found in Sri Lanka that even on long-established and well entrenched savannas the stage could be set for natural reinvasion by native trees by first establishing a plantation of the exotic *Eucalyptus robusta* (Myrtaceae). Although there seems to exist as yet no biological means of controlling *Imperata* (Rao *et al.*, 1971:112–13), a number of mechanical and chemical means can be resorted to (Hubbard *et al.*, 1944; Laurie, 1974; Soerjani, 1970; Tempany, 1951). The possibility of hastening the detoxification of soils polluted with herbicides, although no longer relevant to the herbicidal warfare carried out in Indochina, has been alluded to elsewhere (*see* page 44, note [8]). And although primarily applicable to temperate regions, it can be mentioned here that soil reclamation procedures following saltwater inundation have been described by Dorsman (1947).

IX. *Conclusion*

It should be evident from the foregoing discussion that in a theatre of war the flora, the fauna and the non-living components of the ecosystems of which they are a part are all subject to severe abuse. Indeed, as has been pointed out, such abuse might even be an intentional, systematic and large-scale aspect of a belligerent's military tactics or strategy. It has additionally been seen that such damage is not necessarily confined to its area of application and in fact, that it is difficult thus to confine it. Moreover, the natural recovery of such damage must be measured in years or even decades. And human efforts at shortening these time spans appear not to be feasible except under circumscribed situations.

The question remains of how intense and how widespread military (or other) damage to the natural environment must be before it becomes a matter of grave concern (*see* page 85). On the other hand, any forest damage in Indochina should provoke concern. This is so because of the already precarious position in which one finds the tropical forests of the world.[1]

Notes

[1] There is a large and growing literature on the ecologically or environmentally based problems of the earth's natural resources, unfortunately much of it uncritical. A number of relevant publications can, however, be recommended, among them those by Darling & Milton (1966), Dasmann (1972), Ehrenfeld (1972), Ehrlich & Ehrlich (1972), Falk (1971), W. Thomas *et al.* (1956), Tukey *et al.* (1965) and R. Wagner (1974). The following are important specifically with respect to matters of conservation in the tropics: Bouillenne (1962), Gómez-Pompa *et al.* (1972), Janzen (1973), Richards (1971; 1973) and Talbot & Talbot (1968).

[2] For excellent introductions into the special field of soils (pedology) the reader is referred to Buckman & Brady (1969) and to Lutz & Chandler (1946). For tropical soils one should turn in particular to Mohr *et al.* (1972) and to Nye & Greenland (1960). *See also* Pendleton (1953–1956). Southeast Asian soils have been singled out by Dudal & Moormann (1964) and by Moormann (1961).

Information on soil erosion has been reviewed or compiled by Bennett (1939), Colman (1953), Free & Stuntz (1911), Gaines *et al.* (1938), Katz (1966), Kittredge (1948:XXII–XXIV) and Lutz & Chandler (1946:XIII). Some of the factors peculiar to the tropics which can lead to particularly severe erosion are outlined by Pendleton (1940) and by Nye & Greenland (1960:85–91). Marsh (1864) and V. Carter & Dale (1974) can be especially recommended for their expositions of the fundamental importance of the soil resource to the development and maintenance of human civilization.

[3] The pedological aspects of laterization (induration) are complicated, and are made even more so by a confused nomenclature. Several extensive reviews are available (Alexander & Cady, 1962; Maignien, 1966; Prescott & Pendleton, 1952; Sivarajasingham *et al.*, 1962) as well as a number of briefer articles of note (McNeil, 1964; Pendleton, 1941).

In summary, soils capable of indurating – often referred to as being among the latosols – are rich in sesquioxides of iron and/or aluminium (Fe_2O_3, Al_2O_3) and are thus red through yellowish-red to brownish-red in colour. They have a poorly developed profile (that is, indistinct horizons) and are low in silica (SiO_2) (unless it is in the form of quartz) and in organic matter. They are, moreover, strongly leached and are thus highly deficient in such exchangeable cations as phosphorus, potassium, and calcium. They exhibit an acid reaction. With the protective cover of vegetation stripped away from such soils for an extended period, induration occurs for more or less obscure reasons.

[4] For introductions to the subject of nutrient (biogeochemical) cycles, *see* E. Odum (1971:IV), Spurr & Barnes (1973:VIII), Bormann & Likens (1970), and Jordan *et al.* (1972). Tropical nutrient cycles have been covered by Nye & Greenland (1960:III), Nye (1961), and Jordan & Kline (1972).

[5] Mineral nutrients are washed out of the atmosphere each time it rains or snows and are also deposited rather continuously during clear days as dry fallout. Although its ecological significance is not widely recognized, this form of nutrient accession by an ecosystem appears to be an important one. A growing literature provides data to support this contention (Emanuelsson *et al.*, 1954; Eriksson, 1952; Feth, 1967; Jackson *et al.*, 1973; Likens *et al.*, 1970; Madgwick & Ovington, 1959; Mokma *et al.*, 1972; Voigt, 1960; P. Walker & Costin, 1971; H. Whitehead & Feth, 1964).

[6] A number of synoptic coverages of watershed management and hydrology are available, including those by Colman (1953), Dooge *et al.* (1973), Geiger (1965:360–64), Glesinger (1962:V), Hewlett & Nutter (1969), Hughes (1949–1950:II), Kittredge (1948: XX–XXI), Lull & Reinhart (1972), Penman (1963) and Sopper & Lull (1967). Moreover, pertinent bibliographies have been prepared by Douglass (1972), Gaines *et al.* (1938) and Horton (1973). Most of the relevant research deals with the temperate zones on the one hand and with forest vegetation on the other.

[7] Several reviews of ecological recovery, or succession, can be recommended for one reason or another, among them those by Daubenmire (1968:III), E. Odum (1971:IX), Spurr & Barnes (1973:XIII–XIV) and Toumey & Korstian (1947:XVIII). With respect to the tropics in particular, one can turn to Richards (1952:XVII), Taylor (1957) and L. Williams (1967:163–71). Succession following herbicidal decimation has been described by Hunter & Young (1972) and Young (1974:III).

Animal succession, and especially its relation to plant succession, has been singled out by Beckwith (1954), Diamond (1974), Johnston & Odum (1956), A. Leopold (1933) and A.S. Leopold (1950).

[8] Land reclamation necessitated by strip (opencast) mining has been covered in some detail by Hutnick & Davis (1973). For relevant tropical information, *see* Woudt & Uehara (1961) and Younge & Moomaw (1960). A number of extensive bibliographies on the subject of strip-mine reclamation are also available (Frawley, 1971; Funk, 1962; Kieffer, 1972).

Chapter 7. The implications of environmental disruption by war

Where indicated thus,[1] the reader is referred to the notes on page 89.

I. *Introduction*

Previous chapters have described the principal environmentally disruptive weapons and techniques that were employed by the United States during the Second Indochina War. Estimates were made of the magnitude or intensity of employment of each of these often innovative approaches to counter-guerrilla warfare (chapters 2–5). In each instance an attempt was made to assess the magnitude of environmental disruption. The ecology of disturbance – including the expected character and rapidity of recovery – was explored as well, although in a somewhat more general way (chapter 6).

This concluding chapter examines in brief the extent to which the environmentally disruptive methods pioneered and employed during the Second Indochina War might have been considered sufficiently successful from the military viewpoint to be used again in future wars (section II). It also outlines the several ways being suggested for placing limitations on such war-related environmental disruptions (section III).

II. *Potential for war-related disruption of the environment*

The United States fought a long and costly war to a conclusion it must have found to be an unsatisfactory one. What does this imply for the future regarding the military attractiveness of the environmentally disruptive techniques employed? Detailed military evaluations of the several counterinsurgency methods that have an appreciable environmental impact are beyond the scope of the present work. A brief discussion of this matter is warranted nonetheless, especially in the light of the modern frequency of local (limited) wars (SIPRI, 1969:359–73; Wright, 1965: table 41, Appendix C). First considered is the employment of high-explosive munitions, then that of chemical anti-plant agents and finally that of landclearing tractors.

In analyzing the role of massive bombing in counterinsurgency warfare, the chief historian of the US Strategic Air Command explained that guerrillas are not fought with rifles, but rather are located and then bombed to oblivion (R. Kipp, 1967–1968). Such a tactic requires high levels of technological sophistication and affluence on the part of the user. Moreover, the military utility of massive interdiction (use-denial) bombing carried out over wide rural areas for protracted periods has become a matter of relatively open debate within military circles. One can point to Clelland (1969) or Heavner (1970–1971), both

US Air Force officers, as providing examples of rather half-hearted and circumlocutory attempts at explaining away the "failure" of the US air effort in Indochina. VanderEls (1971), a US Army officer, argues somewhat more boldly and cogently against the effectiveness of interdiction bombing. In fact an analysis of this technique led him to conclude that "massive strategic interdiction fosters the illusion that it will force the enemy to one's desires. It envisions eliminating the evil at its source with all the remote impunity of a surgeon severing a cancerous tumor. It was not a new idea [in Indochina] nor . . . a successful one" (VanderEls, 1971:90). On the other hand, one cannot be so naive as to assume that massive bombing will ever be willingly abrogated by those armed forces of the world capable of carrying it out.

Next there is forest destruction by chemical or mechanical means. The strategic and tactical importance of forests has been recognized throughout the sweep of military history,[1] and the Second Indochina War was certainly no exception. Military evaluations of the use of chemical anti-plant agents as a means of forest destruction have been sufficiently favourable to suggest their future applicability in a diversity of potential operational theatres (Engineers, 1972; Howard, 1972). For example, they are said to have potential military applicability to Cuba, Ethiopia, Korea, Venezuela, western Europe (Engineers, 1972:III; *see* Greenberg, 1972a; Shapley, 1972a), and central Europe (Lyons *et al.*, 1971). Paramilitary herbicide operations (that is, patterned after the military ones) have also been suggested (Nihart, 1970–1971) and perhaps even carried out (Garrett, 1974). It should be added that the technology of chemical anti-plant warfare is widely available and presumably within the financial reach of many armed forces of the world. On the other hand, there appears to exist a rather widespread revulsion against the use of chemical anti-personnel warfare agents, a feeling which encompasses the herbicides used in war to a greater or lesser extent. Proscriptions against chemical weapons *per se* are discussed elsewhere (section III).

Turning next to mechanical means of forest removal, the use of the so-called Rome ploughs has been lauded repeatedly by the military as an aid to counterinsurgency warfare. For example, Major General Robert Ploger, Chief US Army Engineer in Indochina, referred to the Rome-plough operations as "striking", going on to say that "the new land-clearing machinery is an exciting development [which] is changing the face of this jungle-covered country, and eventually may change the face of the war being fought there" (Ploger, 1968:72). There is no difficulty in finding support for his conclusion. Lieutenent Colonel Paul C. Driscoll, commander of the first landclearing battalion formed in Indochina has been quoted as stating that "the B-52 bomber is the battle-ax of this war, and our plow is the scalpel . . . Land clearing in South Vietnam may turn out to be the war's No. 1 [tactical development]" (Engineering News-Record, 1970). A survey of several hundred US Army officers with experience from Indochina was said to have revealed that Rome ploughs were superior for clearing foliage to herbicides, napalm, high-explosive bombs, or felling and burning (Engineers, 1972; *see also* Howard, 1972). Postwar military evaluations are also sprinkled with superlatives (Hay, 1974:87–89; Kerver, 1974; Ploger, 1974:95–104; Rogers, 1974:61–66). And, as with herbicidal agents, mechanized landclearing would appear to be within the technological and financial capabilities of many of the world's armed forces.

Thus it would seem, at least from the military perspective, that various of the environmentally most disruptive weapons and techniques developed and used during the Second Indochina War have a potentially bright future. Any necessary restraints would therefore have to find their basis in moral, ethical or ecological considerations, presumably expressed in the form of legal sanctions of an international nature (section III).

III. *Limiting war-related disruption of the environment*

This section mentions the instruments of international law, both existing and proposed, that appear to have relevance to hostile environmental disruption specifically of the sort which occurred during the Second Indochina War. Those instruments that exist are, in fact, essentially non-existent and proposed ones are few in number.[2]

One might begin the present compilation with the Charter of the United Nations, doing so for the sake of completeness inasmuch as it has the force of treaty law for member nations. This charter establishes that "[all] Members shall settle their international disputes by peaceful means . . . [and] . . . shall refrain in their international relations from the threat or use of force" (United Nations, 1945: Article 2; but *see* Article 51). Adherence to the spirit of that covenant would, *inter alia,* obviate the necessity of further proscriptions against military anti-environmental actions.[3]

The additional existing legal instruments that appear to be relevant to the environmental disruption that occurred during the Second Indochina War are those which deal with the use of chemical weapons. The first of these to be mentioned, since the United States has long been a party to it, is the Annex to Hague Convention IV of 18 October 1907 (SIPRI, 1971–1975:Vol. III:153; Dupuy & Hammerman, 1973:65; Schindler & Toman, 1973:76). Article 23 of this treaty states that "it is especially forbidden . . . to employ poison or poisoned weapons". A second relevant instrument is the Geneva Protocol of 17 June 1925 (Arms Control and Disarmament Agency, 1975:14; SIPRI 1971–1975:Vol. III:155; Schindler & Toman, 1973:110), to which the United States is, however, only a post-war party (Ford, 1975). This well-known arms limitation agreement prohibits "the use in war of asphyxiating, poisonous or other gases, and of all analagous liquids, materials or devices". Adherence to these treaties would seem to preclude the employment of chemical warfare agents of the sort used in Indochina for environmental disruption or any other purposes (*see* pages 24–45 and 53–55). Adherence would, of course, have no bearing on such disruption if carried out by any other means.

With the enumeration of existing documents seemingly exhausted,[3] consideration can next be given to the several proposed legal documents that address themselves to the control of military disruption of the environment. Some of these proposals approach the problem rather broadly whereas others seem limited to instances of intentional disruption. It appears that US Senator Claiborne Pell (1972b; 1973b) is the first to have put forth an explicit proposal of relevance to the present discussion. Attributing his motivation to a desire for improving the human environment and to the great danger from environmental warfare to the "world ecological system", Pell's draft treaty would have ratifying states "prohibit and prevent, at any place, any environmental or geophysical modification activity as a weapon of war . . . [and also] . . . any research or experimentation directed to the development of any such activity" (from Article I). This proposal (which became the "sense" of the US Senate on 11 July 1973) went on to define the operative term "environmental or geophysical modification activity" to signify, *inter alia,* "any weather modification activity which has as a purpose, or has as one of its principal effects, a change in the atmospheric conditions over any part of the earth's surface, including, but not limited to, any activity designed to increase or decrease precipitation" (from Article II). (*See* page 55).

The second such proposal to have been made is the one by Falk (1973:93 – 96) and is meant to define and prohibit environmental warfare or ecocide. Falk's suggestions are a direct outgrowth of the methods employed by the USA during the Second Indochina War. According to him:

> Ecocide means any of the following acts committed with intent to disrupt or destroy, in whole or in part, a human ecosystem:
>
> (a) The use of weapons of mass destruction, whether . . . chemical, or other;
>
> (b) The use of chemical herbicides to defoliate and deforest natural forests for military purposes;
>
> (c) The use of bombs and artillery in such quantity, density, or size as to impair the quality of the soil or to enhance the prospect of diseases dangerous to human beings, animals, or crops;
>
> (d) The use of bulldozing equipment to destroy large tracts of forest or cropland for military purposes;
>
> (e) The use of techniques designed to increase or decrease rainfall or otherwise modify weather as a weapon of war (from Convention Article II).

Falk would, moreover, have states commit themselves

> to refrain from the use of tactics and weapons of war that inflict irreparable harm to the environment or disrupt fundamental ecological relationships; . . . prohibit[ing] in particular:
>
> 1. All efforts to defoliate or destroy forests or crops by means of chemicals or bulldozing;
>
> 2. Any pattern of bombardment that results in extensive craterization of the land or in deep craters that generate health hazards;
>
> 3. Any reliance on weapons of mass destruction of life or any weapons or tactics that are likely to kill or injure large numbers of animals (from Protocol paragraphs 5 – 6).

The Union of Soviet Socialist Republics (1974) was the first nation to put forth a proposed "Convention on the Prohibition of Action to Influence the Environment and Climate for Military and other Purposes Incompatible with the Maintenance of International Security, Human Well-being and Health". The activities prohibited to states ratifying this proposal would include

> those active influences on the surface of the land, . . . the atmosphere or any other elements of the environment that may cause damage by the following means:
>
> (a) introduction into the cloud systems (air masses) of chemical reagents for the purpose of causing precipitation (formation of clouds) and other means of bringing about a redistribution of water resources;
>
> (b) modification of the elements of the weather, climate and the hydrological system on land in any part of the surface of the earth; . . .
>
> (i) modification of the natural state of the rivers, lakes, swamps and other aqueous elements of the land by any methods or means, leading to reduction in the water-level, drying up, flooding, inundation, destruction of hydrotechnical installations or having other harmful consequences;
>
> (j) disturbance of the natural state of the lithosphere, including the land surface, by mechanical, physical or other means, causing erosion, a change in the mechanical structure, desiccation or flooding of the soil, or interference with irrigation or land improvement systems;
>
> (k) the burning of vegetation and other actions leading to a disturbance of the ecology of the vegetable and animal kingdom (from Article II).

Two final items of interest can be mentioned. First, the "Diplomatic Conference on the Reaffirmation and Development of International Humanitarian Law Applicable in Armed Conflicts", meeting in Geneva from February to April 1975, has also been considering matters of the environment (Blix, 1975:126–29, 168–71). Committee III of this Conference, charged with matters concerning the protection of civilians, has made the following relevant proposals:

> It is forbidden to employ methods or means of warfare which are intended or may be expected to cause widespread, long-term, and severe damage to the natural environment (from Article 33)
>
> and
>
> Care shall be taken in warfare to protect the natural environment against widespread, long-term and severe damage. Such care includes a prohibition on the use of methods or means of warfare which are intended or may be expected to cause such damage to the natural environment and thereby to prejudice the health or survival of the population (from Article 48 *bis*).

The most recent and diplomatically most interesting development in the area of limitations on environmental disruption through military activities occurred at the "Conference of the Committee on Disarmament" in Geneva on 21 August 1975. On that date the USA and the USSR presented to this Conference identical, though separate, proposals of a "Draft Convention on the Prohibition of Military or any other Hostile Use of Environmental Modification Techniques" (Union of Soviet Socialist Republics, 1975; United States of America, 1975). With the realization that substantial changes in the environment could be "harmful to human welfare", those states ratifying the proposal would agree "not to engage in military or any other hostile use of environmental modification techniques having widespread, long-lasting or severe effects as the means of destruction, damage or injury" (from Article I). The proposal went on to define the operative term "environmental modification techniques" as, *inter alia*, "any technique for changing – through the deliberate manipulation of natural processes – the dynamics, composition or structure of the earth, including its biota, lithosphere, hydrosphere, and atmosphere . . . so as to cause such effects as . . . an upset in the ecological balance of a region, or changes in weather patterns (clouds, precipitation . . .)" (from Article II).

Each of the several treaty proposals just described has its areas of strength and its shortcomings, both substantive and procedural, but an analysis along these lines is beyond the scope of the present work. However, any evaluation of the substantive aspects of these drafts would have to be made in the light of at least the following considerations: (*a*) hostile *versus* non-hostile military actions, (*b*) incidental *versus* intentional environmental disruptions and – for the latter – whether admitted or not, (*c*) the dimensions of "widespread", (*d*) the duration of "long-lasting" and whether expectation or likelihood suffices in this regard, and (*e*) the magnitude of "severe" and to what extent the criteria of severity are anthropocentric as opposed to natural or ecological *sensu stricto* (section IV).

Despite the indicated trepidations, it is most heartening to observe the growing international interest in the environment within diplomatic circles, and especially within the arms control and disarmament community. The several draft treaties outlined above were motivated only in part by the Second Indochina War. Nevertheless, it appears as if that war would have been pursued rather differently had any of the treaties been in force and respected in letter and spirit.

IV. *Conclusion*

It is axiomatic that warfare is detrimental to the environment (Kristoferson, 1975). To begin with, the preparation for war is detrimental in several ways. It consumes scarce and non-renewable resources, usually on a priority basis. It also generates a variety of pollutants during the manufacture and testing of the war matériel, thereby having an adverse effect on yet other natural resources. The present arms race is particularly wasteful of global resources (Chacko *et al.*, 1972; Myrdal *et al.*, 1972). The present concern, however, is with environmental disruption within an active theatre of war.

Military disruption of the environment during hostilities is pernicious because it spills over both the spatial and temporal boundaries of the attack, because of its partially unpredictable ramifications, and because its impact does not discriminate clearly between combatants and non-combatants. Military disruption of the environment is, however, exceedingly difficult to limit or control by legal instruments. This is because most hostile and many non-hostile military actions result in at least some level of environmental disturbance, whether intended (overtly or covertly) or not, and because of the subsequent difficulties in establishing the magnitude of disruption to be proscribed and the means of determining whether it has, in fact, been exceeded.

A fundamental consideration to ponder is the philosophical basis for a concern over environmental warfare. The several proposed limitations of environmental warfare all appear to be motivated by anthropocentric concerns. The opening statement of Pell's (1972b) draft treaty refers to "human betterment". Falk's (1973) proposal to prevent ecocide defines that term as a disruption of a "human ecosystem" and includes the generation of "health hazards" among his proscriptions. The proposed treaty of the Union of Soviet Socialist Republics (1974) would prohibit actions incompatible with "human well-being and health". The protocol under consideration by the "Diplomatic Conference on the Reaffirmation and Development of International Humanitarian Law Applicable in Armed Conflicts" would prohibit disturbing ecosystems so as "to prejudice the health or survival of the population" (Blix, 1975:168). And the identical treaties proposed to the "Conference of the Committee on Disarmament" by the Union of Soviet Socialist Republics (1975) and the United States of America (1975) are justified on the bases of preventing effects harmful to "human welfare" and of "saving mankind".

But should not living things, and nature as a whole, have some level of immunity in their own right? Is it proper to categorize plants, animals, and their *milieu* in such military terms as "friendly" and "enemy"? Or should they not rather be considered more nearly as innocent and inescapably enmeshed "noncombatant bystanders" to man's martial foibles? What, indeed, is man's fundamental relationship to the land? To what extent can he consider the land in terms of property or real estate and to what extent should he be considering himself as temporary resident and guardian of the land? These and related themes are developed in part by Christopher Stone (1972) in an important essay entitled "Should Trees have Standing? Toward Legal Rights for Natural Objects" which might well be made part of any diplomatic considerations of these problems.

Notes

[1] The military significance of forests in ancient times has been reviewed by Winters (1974) as part of a larger examination of the interrelationship of forests and man. Further sources of information on the strategic significance of forests include Clausewitz (1832–1834:426–27, 530–31), Mammen (1916), Ise (1920) and Dana (1956). Discussions of the tactical significance of forests include those by Schultze (1915), Mammen (1916), Demorlaine (1919) and the US Army (1962).

[2] The literature dealing with arms control and disarmament as it applies to ecology or the environment is quite sparse: Bach (1971), Falk (1973; 1974), Israelyan (1974), Johnstone (1970–1971) and Westing (1971–1972a; 1974c). It might be added here that limitations on environmental warfare are not *sensu stricto* arms control or disarmament measures, but can, of course, be considered to fall into this category of diplomatic endeavours.

International instruments of arms control and disarmament in general have been compiled and annotated a number of times (Arms Control and Disarmament Agency, 1975; Dupuy & Hammerman, 1973; Schindler & Toman, 1973) and there is a useful bibliography on the subject (Rosenblad, 1974). For chemical and biological war (CBW) in particular, *see* SIPRI (1971–1975:Vols. III, IV, V). The US Army (1956) has provided an interpretation of these documents as they are considered to apply to the USA.

[3] Arms control and disarmament endeavours, if not directed toward general and complete disarmament, have traditionally approached the subject from one of three directions: the focus has been on particular weapons (for example, chemical ones), or on man *per se* (either as combatant or non-combatant), or on geographic regions of one sort or another. Any of these approaches can have ancillary environmental benefits. The case of chemical weapons is discussed in the text (*see* page 85). An incident from World War II provides an example for the anthropocentric approach. The pillaging of Polish forests by the Germans during that war was classed as a war crime by the United Nations War Crimes Commission (1948:496) at Nuremberg, in Case No. 7150. It was considered to be a violation of Article 47 of the Annex to Hague Convention IV of 18 October 1907 (Schindler & Toman, 1973:83). The relevance of regional controls such as the Antarctic Treaty of 1959, the Outer Space Treaty of 1967, or the Sea-Bed Treaty of 1971 (Arms Control and Disarmament Agency, 1975) is that they provide for the protection of a region in the same sense that a wildlife refuge or sanctuary does. Not only is protection thereby provided for the involved ecosystems with their living and non-living components, but also perhaps for some crucial link in the greater cycles of nature that serve to maintain the biosphere, or worldwide ecosystem. Under consideration at the "Diplomatic Conference on the Reaffirmation and Development of International Humanitarian Law Applicable in Armed Conflicts" is Article 48 *ter* on the "Protection of Nature Reserves" which states that "nature reserves with adequate markings and boundaries declared as such to the adversary shall be protected and respected except when such reserves are used specifically for military purposes" (Mahony, 1975; Blix, 1975:Appendix 19). In connection with this, South Viet-Nam's only two national parks are said to have suffered serious herbicidal and other war damage (Kopp, 1972).

References

References to publications in the text provide information sufficient to locate the full bibliographic citation in the alphabetical listing of references, that is, author and date. Separate authors who share a surname are distinguished in the text by the initials of their given names. Different publications by the same author that were published during the same year are distinguished by the arbitrary assignment of a series of lower case letters appended to the year of publication. Additional numbers are in some instances provided in the text immediately following the year of publication (being separated from it by a colon). These refer to specific portions or locations within the publication. Roman numerals signify chapters and Arabic numerals signify pages. The occasional deviations from this last convention have been necessitated by the oddities of pagination in the publication being cited and should become clear when the original is consulted.

Adams, J.B., 1960, Effects of spraying 2,4-D amine on coccinellid larvae. *Canadian Journal of Zoology*, Ottawa, **38**, pp. 285–88.

Adams, J.B. & Drew, M.E., 1965, Grain aphids in New Brunswick. III. Aphid populations in herbicide-treated oat fields. *Canadian Journal of Zoology*, Ottawa, **43**, pp. 789–94.

Adams, J.B. & Drew, M.E., 1969, Grain aphids in New Brunswick. IV. Effects of malathion and 2,4-D amine on aphid populations and on yields of oats and barley. *Canadian Journal of Zoology*, Ottawa, **47**, pp. 423–26.

Adams, R.A., 1971, Anchor chain land clearing. *Military Engineer*, Washington, **63**, p. 259.

Adams, S., 1975, Vietnam cover-up: playing war with numbers. *Harper's*, New York, **250** (1500), pp. 41–44, 62–73.

Agricultural Research Service, US, 1969, *Chemical Control of Brush and Trees*. US Department of Agriculture, Washington, Farmers' Bulletin No. 2158, 23 pp.

Ahlgren, I.F. & Ahlgren, C.E., 1960, Ecological effects of forest fires. *Botanical Review*, New York, **26**, pp. 483–533.

Air Force, US Department of the, 1971, *Joint Munitions Effectiveness Manual (Air to Surface): Defoliants (JMEM)*. US Department of the Air Force, Washington, Technical Handbook No. 61A1-1-1-4, [*ca.* 35] pp.

Air Force, US Department of the, 1974, *Revised Draft Environmental Statement on Disposition of Orange Herbicide by Incineration*. US Department of the Air Force, Washington, Publication No. AF-ES-72-2D (1), 125+[403] pp.

Air University Quarterly Review, 1953–1954, Attack on the irrigation dams in North Korea. *Air University Quarterly Review*, Maxwell Air Force Base, Alabama, **6**(4), pp. 40–61.

Akamine, E.K., 1950–1951, Persistence of 2,4-D toxicity in Hawaiian soils. *Botanical Gazette*, Chicago, **112**, pp. 312–19.

Aldrich, R.J., 1953, Herbicides: residues in soil. *Journal of Agricultural and Food Chemistry*, Washington, **1**, pp. 257–60.

Alexander, L.T. & Cady, J.G., 1962, *Genesis and Hardening of Laterite in Soils*. US Department of Agriculture, Washington, Technical Bulletin No. 1282, 90 pp. + 8 pl.

Allee, W.C. & Schmidt, K.P., 1951, *Ecological Animal Geography*, 2nd ed. John Wiley, New York, 715 pp.
Allen, D.L., 1962, *Our Wildlife Legacy*, rev. ed. Funk and Wagnalls, New York, 422 pp. + pl.
American Chemical Society, 1955, *Pesticides in Tropical Agriculture*. American Chemical Society, Washington, Advances in Chemistry Series No. 13, 102 pp.
Anderson, H.W., Duffy, P.D. & Yamamoto, T., 1966, *Rainfall and Streamflow from Small Tree-covered and Fern-covered and Burned Watersheds in Hawaii*. US Forest Service, Washington, Research Paper No. PSW-34, 10 pp.
Anderson, J., 1971, Air Force turns rainmaker in Laos. *Washington Post*, 18 March 1971, p. F7.
Anglemyer, M., Gee, J.G. & Koll, M.J., Jr., 1969, *Selected Bibliography: Lower Mekong Basin*. United Nations, New York, 2 vols.
Arend, J.L., 1941, Infiltration rates of forest soils in the Missouri Ozarks as affected by woods burning and litter removal. *Journal of Forestry*, Washington, **39**, pp. 726–28.
Arms Control and Disarmament Agency, US, 1975, *Arms Control and Disarmament Agreements: Texts and History of Negotiations*. US Arms Control and Disarmament Agency, Washington, Publication No. 77, 159 pp.
Army, US Department of the, 1950, *Field Artillery Gunnery*, US Department of the Army, Washington, Field Manual No. 6-40, 495 pp. + 1 fig.
Army, US Department of the, 1956, *Law of Land Warfare*. US Department of the Army, Washington, Field Manual No. 27–10, 236 pp.
Army, US Department of the, 1962, *Barriers and Denial Operations*. US Department of the Army, Washington, Field Manual No. 31–10, 128 pp.
Army, US Department of the, 1967a, *Artillery Ammunition: Guns, Howitzers, Mortars and Recoilless Rifles*. US Department of the Army, Washington, Technical Manual No. 9-1300-203, (var. pp.).
Army, US Department of the, 1967b, *Chemical Reference Handbook*. US Department of the Army, Washington, Field Manual No. 3–8, 132 pp.
Army, US Department of the, 1969, *Employment of Riot Control Agents, Flame, Smoke, Antiplant Agents, and Personnel Detectors in Counterguerrilla Operations*. US Department of the Army, Washington, Training Circular No. 3–16, 85 pp.
Army, US Department of the, 1972, *Peninsular Southeast Asia: A Bibliographic Survey of Literature*. US Department of the Army, Washington, Pamphlet No. 550–14, 424 pp. + 17 maps.
Army, Navy, & Air Force, US Departments of the, 1966, *Bombs and Bomb Components*. US Department of the Army, Washington, Technical Manual No. 9-1325-200, [309] pp.
Arnett, P., 1969, After 2 years in Mekong delta, U.S. goal is elusive. *New York Times*, 15 April 1969, p. 12.
Arnold, W.R., Santelmann, P.W. & Lynd, J.Q., 1966, Picloram and 2,4-D effects with *Aspergillus niger* proliferation. *Weeds*, Urbana, Illinois, **14**, pp. 89–90.
Ashton, F.M. & Crafts, A.S., 1973, *Mode of Action of Herbicides*. John Wiley, New York, 504 pp.
Associated Press, 1971a, U.S. aerosol bomb: mist that kills. *San Francisco Chronicle*, 28 May 1971, p. 18.
Associated Press, 1971b, U.S. planes start jungle fires to aid besieged Vietnam base. *New York Times*, 10 April 1971, p. 7.
Associated Press, 1975, U.S. war casualties. *New York Times*, 1 May 1975, p. 20.
Audus, L.J., 1970, Action of herbicides and pesticides on the microflora. *Mededelingen Faculteit Landbouwwetenschappen*, Ghent, **35**, pp. 465–92.
Audy, J.R., 1948, Some ecological effects of deforestation and settlement. *Malayan Nature Journal*, Kuala Lumpur, **4**(4), pp. 178–89 + 3 figs.

Bach, Pham Van., 1971, Law and the use of chemical warfare in Vietnam. *Scientific World*, London, **15**(6), pp. 12–14.
Batchelder, R.B. & Hirt, H.F., 1966, *Fire in Tropical Forests and Grasslands*. US Army Laboratories, Natick, Massachusetts, Earth Sciences Division, Report No. ES-23, 380 pp.
Batjer, L.P. & Benson, N.R., 1958, Effect of metal chelates in overcoming arsenic toxicity to peach trees. *Proceedings of the American Horticultural Society*, Washington, **72**, pp. 74–78.
Beckwith, S.L., 1954, Ecological succession on abandoned farm lands and its relationship to wildlife management. *Ecological Monographs*, US, **24**, pp. 349–76.
Beek, F.t., Bokhorst, R., Plas, M.v.d., Olsthoorn, K., Vergragt, P. & Zwan, G.v.d., 1973, [Dioxin: a dangerous contaminant in a much used herbicide.] (In Dutch) *Chemisch Weekblad*, The Hague, **69**(23), pp. 5, 7.
Bennett, H.H., 1939, *Soil Conservation*. McGraw-Hill, New York, 993 pp.
Bentley, J.R., Conrad, C.E. & Schimke, H.E., 1971, *Burning Trials in Shrubby Vegetation Desiccated with Herbicides*. US Forest Service, Washington, Research Note No. PSW-241, 9 pp.
Bethlahmy, N., 1974, More streamflow after a bark beetle epidemic. *Journal of Hydrology*, Amsterdam, **23**, pp. 185–89.
Bingham, S.W., 1973, *Improving Water Quality by Removal of Pesticide Pollutants with Aquatic Plants*. Virginia Polytechnic Institute Water Resource Research Center, Blacksburg, Virginia, Bulletin No. 58, 94 pp.
Björklund, N.-E. & Erne, K., 1966, Toxicological studies of phenoxyacetic herbicides in animals. *Acta Veterinaria Scandinavica*, **7**, pp. 364–90.
Björklund, N.-E. & Erne, K., 1971, Phenoxy-acid-induced renal changes in the chicken. I. Ultrastructure. *Acta Veterinaria Scandinavica*, **12**, pp. 243–56.
Björnerstedt, R. *et al.*, 1973, *Napalm and Other Incendiary Weapons and All Aspects of their Possible Use*. United Nations, New York, 63 pp.
Blackman, G.E., Fryer, J.D., Lang, A. & Newton, M., 1974, *Effects of Herbicides in South Vietnam. B[3]. Persistence and Disappearance of Herbicides in Tropical Soils*. US National Academy of Sciences, Washington, 59 pp.
Blair, E.H., (ed.), 1973, *Chlorodioxins: Origin and Fate*. American Chemical Society, Washington, Advances in Chemistry Series No. 120, 141 pp.
Blix, H., (ed.), 1975, *[Geneva 1975: Diplomatic Conference on Laws of War.]* (In Swedish) Royal Ministry for Foreign Affairs, Stockholm, 215 pp. + 20 apps.
Blumenfeld, S. & Meselson, M., 1971, Military value and political implications of the use of riot control agents in warfare. In: Alexander, A.S. *et al.*, *Control of chemical and biological weapons*. Carnegie Endowment for International Peace, New York, 130 pp., pp. 64–93.
Blumenthal, R., 1969, U.S. now uses tear gas as routine war weapon. *New York Times*, 6 December 1969, p. 3.
Bollen, W.B., 1961, Interactions between pesticides and soil microorganisms. *Annual Reviews of Microbiology*, **15**, pp. 69–92.
Bollen, W.B., Norris, L.A. & Stowers, K.L., 1974, Effect of cacodylic acid and MSMA on microbes in forest floor and soil. *Weed Science*, Urbana, Illinois, **22**, pp. 557–62.
Bormann, F.H. & Likens, G.E., 1970, Nutrient cycles of an ecosystem. *Scientific American*, New York, **223**(4), pp. 92–101, 144.
Bormann, F.H., Likens, G.E. & Eaton, J.S., 1969, Biotic regulation of particulate and solution losses from a forest ecosystem. *BioScience*, Washington, **19**, pp. 600–10.
Bormann, F.H., Likens, G.E., Fisher, D.W. & Pierce, R.S., 1968, Nutrient loss accelerated by clear-cutting of a forest ecosystem. *Science*, Washington, **159**, pp. 882–84.
Bormann, F.H., Likens, G.E., Siccama, T.G., Pierce, R.S. & Eaton, J.S., 1974, Export of nutrients and recovery of stable conditions following deforestation at Hubbard

Brook. *Ecological Monographs*, US, **44**, pp. 255–77.
Bouillenne, R., 1962, Man, the destroying biotype. *Science*, Washington, **135**, pp. 706–12.
Bovey, R.W., Dowler, C.C. & Diaz-Colon, J.D., 1969a, Response of tropical vegetation to herbicides. *Weed Science*, Urbana, Illinois, **17**, pp. 285–90.
Bovey, R.W., Dowler, C.C. & Merkle, M.G., 1969b, Persistence and movement of picloram in Texas and Puerto Rican soils. *Pesticides Monitoring Journal*, Washington, **3**, pp. 177–81.
Bovey, R.W. & Miller, F.R., 1969, Effect of activated carbon on the phytotoxicity of herbicides in a tropical soil. *Weed Science*, Urbana, Illinois, **17**, pp. 189–92.
Bovey, R.W., Miller, F.R. & Diaz-Colon, J., 1968, Growth of crops in soils after herbicidal treatments for brush control in the tropics. *Agronomy Journal*, Madison, Wisconsin, **60**, pp. 678–79.
Bovey, R.W. & Scifres, C.J., 1971, *Residual Characteristics of Picloram in Grassland Ecosystems*. Texas Agricultural and Mechanical University Agricultural Experiment Station, College Station, Texas, Bulletin No. B-1111, 24 pp.
Braman, R.S. & Foreback, C.C., 1973, Methylated forms of arsenic in the environment. *Science*, Washington, **182**, pp. 1247–49.
Breazeale, F.W. & Camper, N.D., 1970, Bacterial, fungal, and actinomycete populations in soils receiving repeated applications of 2,4-dichlorophenoxyacetic acid and trifluralin. *Applied Microbiology*, Bethesda, Maryland, **19**, pp. 379–80.
Broido, A., 1963, Effects of fire on major ecosystems. In: Woodwell, G.M., (ed.), *Ecological Effects of Nuclear War*. Brookhaven National Laboratory, Upton, New York, Publication No. 917, 72 pp., pp. 11–19.
Brouillard, K.D., 1970, *Fishery Development Survey: South Vietnam*. US Agency for International Development, Saigon, 41 pp.
Brown, A.A. & Davis, K.P., 1973, *Forest Fire: Control and Use*, 2nd ed. McGraw-Hill, New York, 686 pp.
Brun, W.A., Cruzado, H.J. & Muzik, T.J., 1961, Chemical defoliation and desiccation of tropical woody plants. *Tropical Agriculture*, Trinidad, **38**(1), pp. 69–81.
Bryson, R.A., 1974, Perspective on climatic change. *Science*, Washington, **184**, pp. 753–60.
Bryson, R.A. & Baerreis, D.A., 1967, Possibilities of major climatic modification and their implications: northwest India, a case for study. *Bulletin of the American Meteorological Society*, Boston, **48**, pp. 136–42.
Buckman, H.O. & Brady, N.C., 1969, *Nature and Property of Soils*, 7th ed. Macmillan, New York, 653 pp. + 1 fig.
Budowski, G., 1956, Tropical savannas, a sequence of forest felling and repeated burnings. *Turrialba*, Costa Rica, **6**(1–2), pp. 23–33.
Buffam, P.E., Lister, C.K., Stevens, R.E. & Frye, R.H., 1973, Fall cacodylic acid treatments to produce lethal traps for spruce beetles. *Environmental Entomology*, US, **2**, pp. 259–62.
Butler, P.A., 1965, Effects of herbicides on estuarine fauna. *Proceedings of the Southern Weed Conference*, US, **18**, pp. 576–80.
Canatsey, J.D., 1968, Bigger bite for bulldozers. *Military Engineer*, Washington, **60**, pp. 256–57.
Cannon, H.W., (ed.), 1971, *Investigation Into Electronic Battlefield Program*. US Senate, Washington, Committee on Armed Services, 221 pp.
Carlson, A.W., 1974, *Bibliography of the Geographical Literature on Southeast Asia, 1920–1972*. Council of Planning Librarians, Monticello, Illinois, Exchange Bibliography No. 598–599–600, 127 pp.
Carson, R., 1962, *Silent Spring*. Houghton Mifflin, Boston, 368 pp.
Carter, C.D., Kimbrough, R.D., Liddle, J.A., Cline, R.E., Zack, M.M., Jr., Barthel,

W.F., Koehler, R.E. & Phillips, P.E., 1975, Tetrachlorodibenzodioxin: an accidental poisoning episode in horse arenas. *Science,* Washington, **188**, pp. 738–40.
Carter, V.G. & Dale, T., 1974, *Topsoil and Civilization,* rev. ed. University of Oklahoma Press, Norman, Oklahoma, 292 pp. + pl.
Caterpillar Tractor Company, 1970, *Land Clearing.* Caterpillar Tractor Company, Peoria, Illinois, 107 pp.
Chacko, M.E. *et al.,* 1972, *Economic and Social Consequences of the Arms Race and of Military Expenditures.* United Nations, New York, 51 pp.
Challenger, F., 1945, Biological methylation. *Chemical Reviews,* Washington, **36**, pp. 315–61.
Changnon, S.A., Jr., 1968, LaPorte weather anomaly: fact or fiction? *Bulletin of the American Meteorological Society,* Boston, **49**, pp. 4–11.
Chansler, J.F., Cahill, D.B. & Stevens, R.E., 1970, *Cacodylic Acid Field Tested for Control of Mountain Pine Beetles in Ponderosa Pine.* US Forest Service, Washington, Research Note No. RM-161, 3 pp.
Chapman, D.W., 1962, Effects of logging upon fish resources of the west coast. *Journal of Forestry,* Washington, **60**, pp. 533–37.
Chapman, V.J., 1970, Mangrove phytosociology. *Tropical Ecology,* Varanasi, India, **11**(1), pp. 1–19.
Chevalier, A., 1948, [Biogeography and ecology of the dense ombrophilous forest of the Ivory Coast.] (In French) *Revue Internationale de Botanique Appliquée & d'Agriculture Tropicale,* Paris, **28**, pp. 101–15.
Clausewitz, [C.]v., 1832–1834, *On War* [translated from the German by O.J.M. Jolles]. Infantry Journal Press, Washington, 641 pp., 1950.
Clelland, D., 1969, Air interdiction: its changing conditions. *Air Force and Space Digest,* Washington, **52**(6), pp. 52–56.
Cockrell, R.A., 1971, Side effect of tear gas. *BioScience,* Washington, **21**, p. 778.
Cohn, V., 1972, Weather war: a gathering storm. *Washington Post,* 2 July 1972, pp. C1–C2.
Colman, E., 1953, *Vegetation and Watershed Management: An Appraisal of Vegetation Management in Relation to Water Supply, Flood Control, and Soil Erosion.* Ronald Press, New York, 412 pp. + maps.
Commun, R., 1961, [Chemical warfare against weeds in overseas countries.] (In French) *Agronomie Tropicale,* Paris, **16**, pp. 445–69.
Condon, P.A., 1968, *Toxicity of Herbicides to Mammals, Aquatic Life, Soil Microorganisms, Beneficial Insects and Cultivated Plants, 1950–65: A List of Selected References.* US National Agricultural Library, Bethesda, Maryland, List No. 87, 161 pp.
Constable, J. & Meselson, M., 1971, Ecological impact of large scale defoliation in Vietnam. *Sierra Club Bulletin,* San Francisco, **56**(4), pp. 4–9.
Cook, R.E., 1970–1971, 'Completely destroyed'. *Yale Alumni Magazine,* New Haven, **34**(8), pp. 26–29.
Cooper, C.F., 1961, Ecology of fire. *Scientific American,* New York, **204**(4), pp. 150–60, 207.
Cooper, C.F. & Jolly, W.C., 1969, *Ecological Effects of Weather Modification: A Problem Analysis.* University of Michigan Department of Resource Planning and Conservation, Ann Arbor, Michigan, 160 pp.
Cooper, C.F. & Jolly, W.C., 1970, Ecological effects of silver iodide and other weather modification agents: a review. *Water Resources Research,* Washington, **6**, pp. 88–98.
Cope, O.B., 1966, Contamination of the freshwater ecosystem by pesticides. *Journal of Applied Ecology,* Oxford, **3**(Supplement), pp. 33–44.
Cope, O.B., Wood, E.M. & Wallen, G.H., 1970, Some chronic effects of 2,4-D on the bluegill *(Lepomis macrochirus). Transactions of the American Fisheries Society,* **99**, pp. 1–12.

Corbett, J.R., 1974, *Biochemical Mode of Action of Pesticides.* Academic Press, New York, 330 pp.
Cox, D.P. & Alexander, M., 1973, Production of trimethylarsine gas from various arsenic compounds by three sewage fungi. *Bulletin of Environmental Contamination and Toxicology,* West Berlin, **9**, pp. 84–88.
Crossland, J. & Shea, K.P., 1973, Hazards of impurities. *Environment,* Saint Louis, **15**(5), pp. 35–38.
Cullimore, D.R., 1971, Interaction between herbicides and soil microorganisms. *Residue Reviews,* **35**, pp. 65–80.
Cunningham, R.K., 1963, Effect of clearing a tropical forest soil. *Journal of Soil Science,* UK, **14**, pp. 334–45 + 1 pl.
Curry-Lindahl, K., 1972, *Let Them Live: A Worldwide Survey of Animals Threatened with Extinction.* Morrow, New York, 394 pp.
Cushwa, C.T., 1968, *Fire: A Summary of Literature in the United States From the Mid-1920's to 1966.* US Forest Service Southeastern Forest Experiment Station, Asheville, North Carolina, 117 pp.
DaLage, C. & Alnot, M.-O., 1973, [Teratogenic effects of certain pesticides on the embryos of birds and mammals.] (In French) *Economie et Médecine Animales,* Paris, **14**(3), pp. 141–50.
Dalgaard-Mikkelsen, S. & Poulsen, E., 1962, Toxicology of herbicides. *Pharmacological Reviews,* Baltimore, **14**, pp. 225–50.
Dana, S.T., 1956, *Forest and Range Policy: Its Development in the United States.* McGraw-Hill, New York, 455 pp.
Danielson, L.L. *et al.,* 1969, *Suggested Guide for Weed Control 1969.* US Department of Agriculture, Washington, Agriculture Handbook No. 332, 70 pp.
Darcourt, P., 1971, When floods and typhoons hit North Vietnam. *U.S. News and World Report,* Washington, **71**(20), p. 46.
Darling, F.F. & Milton, J.P., (eds.), 1966, *Future Environments of North America.* Natural History Press, Garden City, New York, 770 pp.
Darrow, R.A., Frank, J.R., Martin, J.W., Demaree, K.D. & Creager, R.A., 1971, *Field Evaluation of Desiccants and Herbicide Mixtures as Rapid Defoliants.* US Army Fort Detrick, Frederick, Maryland, Technical Report No. 114, 55 pp.
Darrow, R.A., Irish, K.R. & Minarik, C.E., 1969, Herbicides Used in Southeast Asia. US Army Fort Detrick, Frederick, Maryland, Plant Sciences Laboratories Technical Report No. SAOQ-TR-69-11078, 60 pp.
Dasmann, R.F., 1972, *Environmental Conservation,* 3rd ed. John Wiley, New York, 473 pp.
Daubenmire, R., 1968, *Plant Communities: A Textbook of Plant Synecology.* Harper and Row, New York, 300 pp.
Davis, G.M., 1973–1974, Defoliation in Vietnam: assessing the damage. *Frontiers,* Philadelphia, **38**(2), pp. 18–23.
Davis, G.M., 1974, *Effects of Herbicides in South Vietnam. B[6]. Mollusks as Indicators of the Effects of Herbicides on Mangroves in South Vietnam.* US National Academy of Sciences, Washington, 29 pp.
Davis, J.H., Jr., 1939–1940, Ecology and geologic rôle of mangroves in Florida. *Papers of the Tortugas Laboratory, Washington,* **32**, pp. 303–412 + 12 pl. + 2 maps.
Dävring, L. & Sunner, M., 1971, Cytogenetic effects of 2,4,5-trichlorophenoxyacetic acid on oogenesis and early embryogenesis in *Drosophila melanogaster. Hereditas,* Lund, **68**, pp. 115–22.
Defense, US Department of, 1968, *United States–Vietnam relations, 1945–1967.* US House of Representatives, Washington, Committee on Armed Services, 12 vols., 1971.
Dellums, R.V., (ed.), 1972, Escalation, American options, and President Nixon's war

moves. *U.S. Congressional Record, Washington,* **118**, pp. 16748–836.
Delvaux, E.L., Verstraete, J. & Hautfenne, A., 1975, [Polychlorodibenzo-*p*-dioxins.] (In French) *Toxicology,* Amsterdam, **3**, pp. 187–206.
Demorlaine, J., 1919, [Strategic importance of forests in war.] (In French) *Revue des Eaux et Fôrets,* Paris, **57**, pp. 25–30.
Desowitz, R.S., Berman, S.J., Gubler, D.J., Harinasuta, C., Guptavanij, P. & Vasuvat, C., 1974, *Effects of Herbicides in South Vietnam. B[7]. Epidemiological-ecological Effects: Studies on Intact and Deforested Mangrove Ecosystems.* US National Academy of Sciences, Washington, 54 pp.
DeVaney, T.E., 1968, *Chemical Vegetation Control Manual for Fish and Wildlife Management Programs.* US Bureau of Sport Fisheries and Wildlife, Washington, Resource Publication No. 48, 42 pp.
Diamond, J.M., 1974, Colonization of exploded volcanic islands by birds: the supertramp strategy. *Science,* Washington, **184**, pp. 803–806.
D'Olier, F. *et al.,* 1947, *Index to Records of the United States Strategic Bombing Survey.* US Strategic Bombing Survey, Washington, 317 pp.
Dooge, J.C.I., Costin, A.B. & Finkel, H.J., 1973, *Man's Influence on the Hydrological Cycle.* Food and Agriculture Organization of the United Nations, Rome, Irrigation and Drainage Paper No. 17, 71 pp. + pl.
Dorsman, C., 1947, [Damage to horticultural crops from inundation with seawater.] (In Dutch with English summary) *Tijdschrift over Plantenziekten,* Wageningen, **53**(3), pp. 65–86.
Douglass, J.E., 1972, *Annotated Bibliography of Publications on Watershed Management by the Southeastern Forest Experiment Station, 1928–1970.* US Forest Service, Washington, Research Paper No. SE-93, 47 pp.
Douglass, J.E. & Swank, W.T., 1972, *Streamflow Modification Through Management of Eastern Forests.* US Forest Service, Washington, Research Paper No. SE-94, 15 pp.
Dowler, C.C., Forestier, W. & Tschirley, F.H., 1968, Effect and persistence of herbicides applied to soil in Puerto Rican forests. *Weed Science,* Urbana, Illinois, **16**, pp. 45–50.
Dowler, C.C. & Tschirley, F.H., 1970, Evaluation of herbicides applied to foliage of four tropical woody species. *Journal of Agriculture of the University of Puerto Rico,* Rio Piedras, **54**, pp. 676–82.
Draper, S.E., 1971, Land Clearing in the Delta, Vietnam. *Military Engineer,* Washington, **63**, pp. 257–59.
Drendel, L., 1968, *Air War in Vietnam.* Arco, New York, 96 pp.
Drew, W.B., 1974a, *Effects of Herbicides in South Vietnam. B[8]. Ecological Role of Bamboos in Relation to the Military Use of Herbicides on Forests of South Vietnam.* US National Academy of Sciences, Washington, 14 pp.
Drew, W.B., 1974b, *Effects of Herbicides in South Vietnam. B[9]. Ecological Role of the Fern* (Acrostichum aureum) *in Sprayed and Unsprayed Mangrove Forests.* US National Academy of Sciences, Washington, 13 pp.
Dubey, H.D., 1969, Effect of picloram, diuron, ametryne, and prometryne on nitrification in some tropical soils. *Soil Science Society of America Proceedings,* Madison, Wisconsin, **33**, pp. 893–96.
Dudal, R. & Moormann, F.R., 1964, Major soils of south-east Asia: their characteristics, distribution, use and agricultural potential. *Journal of Tropical Geography,* Kuala Lumpur, **18**, pp. 54–80.
Duffett, J., (ed.), 1968, *Against the Crime of Silence: Proceedings of the Russell International War Crimes Tribunal: Stockholm, Copenhagen.* O'Hare Books, Flanders, New Jersey, 662 pp.
Dupuy, T.N. & Hammerman, G.M., (eds.), 1973, *Documentary History of Arms Control and Disarmament.* R.R. Bowker, New York, 629 pp.

Ehman, P.J., 1965, Effect of arsenical build-up in the soil on subsequent growth and residue content of crops. *Proceedings of the Southern Weed Conference,* US, **18**, pp. 685–87.

Ehrenfeld, D.W., 1972, *Conserving Life on Earth.* Oxford University Press, New York, 360 pp.

Ehrlich, P.R. & Ehrlich, A.H., 1972, *Population, Resources, Environment: Issues in Human Ecology,* 2nd ed. W.H. Freeman, San Francisco, 509 pp.

Ellis, M.M., 1936, Erosion silt as a factor in aquatic environments. *Ecology,* US, **17**, pp. 29–42.

Emanuelsson, A., Eriksson, E. & Egnér, H., 1954, Composition of atmospheric precipitation in Sweden. *Tellus,* Stockholm, **6**, pp. 261–67.

Emmons, D.M., 1971–1972, Theories of increased rainfall and the Timber Culture Act of 1873. *Forest History,* Santa Cruz, California, **15**(3), pp. 6–14.

Engineer Agency for Resources Inventories & Tennessee Valley Authority, 1968, *Atlas of Physical, Economic and Social Resources of the Lower Mekong Basin.* United Nations, New York, 257 pp.

Engineering News-Record, 1970, Land clearing emerges as a top tactic of the war. *Engineering News-Record,* New York, **184**(3), p. 27.

Engineers, US Army Corps of, 1972, *Herbicides and Military Operations.* US Army Corps of Engineers, Washington, 3 vols. (27+[141]+98 pp.)

Enthoven, A.C. & Smith, K.W., 1971, *How Much is Enough? Shaping the Defense Program, 1961–1969.* Harper & Row, New York, 364 pp.

Epstein, S.S., 1970, Family likeness. *Environment,* Saint Louis, **12**(6), pp. 16–25.

Eriksson, E., 1952, Composition of atmospheric precipitation. I. Nitrogen compounds. II. Sulfur, chloride, iodine compounds. Bibliography. *Tellus,* Stockholm, **4**, pp. 215–32, 280–303.

Erne, K., 1973, Weed-killers and wildlife. *Proceedings of the International Congress of Game Biology,* Stockholm, **11**, pp. 415–22.

Erne, K., 1974, [Herbicides and wildlife: several recent results.] (In German) *Zeitschrift für Jagdwissenschaft,* Hamburg, **20**, pp. 68–70.

Erne, K. & Hartman, U.v., 1973, [Phenoxy herbicide residues in forest berries and mushrooms.] (In Swedish) *Vår Föda,* Stockholm, **25**, pp. 146–54.

Estes, J.A. & Palmisano, J.F., 1974, Sea otters: their role in structuring nearshore communities. *Science,* Washington, **185**, pp. 1058–60.

Fabian, D.R., 1970, Bunker-busting operation. *Military Engineer,* Washington, **62**, pp. 102–103.

Falk, R.A., 1971, *This Endangered Planet: Prospects and Proposals for Human Survival.* Random House, New York, 497 pp.

Falk, R.A., 1973, Environmental warfare and ecocide. *Bulletin of Peace Proposals,* Oslo, **4**, pp. 1–17.

Falk, R.A., 1974, Law and responsibility in warfare: the Vietnam experience. *Instant Research on Peace and Violence,* Tampere, Finland, **4**, pp. 1–14.

Feth, J.H., 1967, *Chemical Characteristics of Bulk Precipitation in the Mojave Desert Region, California,* US Geological Survey, Washington, Professional Paper No. 575-C, 251 pp., pp. C222–C227.

Fisher, M.L., 1967, *Cambodia: An Annotated Bibliography of its History, Geography, Politics and Economy Since 1954.* MIT Center for International Studies, Cambridge, Massachusetts, Publication No. C/67–17, 66 pp.

FitzGerald, F., 1972, *Fire in the Lake: The Vietnamese and the Americans in Vietnam.* Little Brown & Company, Boston, 491 pp.

Fletcher, W.W. & Raymond, J.C., 1956, Toxicity and breakdown of 'hormone' herbicides. *Nature,* London, **178**, pp. 151–52.

Foisie, J., 1970, Thriftiness, ingenuity: U.S. innovations used to push Vietnam war. *Los Angeles Times*, 1 June 1970, Part I, p. 20.
Foldesy, R.G., Mineo, L. & Majumdar, S.K., 1972, Effect of 2,4,5-trichlorophenoxyacetic acid on two nitrogen fixing bacteria. *Proceedings of the Pennsylvania Academy of Science*, **46**, pp. 23–24.
Ford, G.R., 1975, Geneva Protocol of 1925 and Biological Weapons Convention. *U.S. Department of State Bulletin*, Washington, **72**, pp. 576–77.
Forman, O.L. & Longacre, D.W., 1969–1970, Fire potential increased by weed killers. *Fire Control Notes*, Washington, **31**(3), pp. 11–12.
Fox, C.J.S., 1964, Effects of five herbicides on the numbers of certain invertebrate animals in grassland soil. *Canadian Journal of Plant Science*, Ottawa, **44**, pp. 405–409.
Fox, G.W., 1970, *Pesticides and Ecosystems*. US National Library of Medicine, Bethesda, Maryland, Literature Search No. 70–39, 24 pp.
Frankenberg, L. & Sörbo, B., 1973, Formation of cyanide from *o*-chlorobenzylidene malononitrile and its toxicological significance. *Archiv für Toxikologie*, West Berlin, **31**, pp. 99–108.
Frawley, M.L., 1971, *Surface Mined Areas: Control and Reclamation of Environmental Damage: A Bibliography*. US Department of the Interior, Washington, Office of Library Services Bibliography Series No. 27, 63 pp.
Free, E.E. & Stuntz, S.C., 1911, *Movement of Soil Material by the Wind, with a Bibliography of Eolian Geology*. US Department of Agriculture, Washington, Bureau of Soils Bulletin No. 68, 272 pp. + 5 pl.
French, R.W. & Callender, G.R., 1962, Ballistic characteristics of wounding agents. In: Beyer, J.C., (ed.), *Wound Ballistics*. US Department of the Army, Washington, Office of the Surgeon General, 883 pp., pp. 91–141.
Frodigh, R.J., Hastings, A.D., Jr., Niedringhaus, T.E. & Planalp, J.M., 1969, *Mainland Southeast Asia: A Folio of Thematic Maps for Military Users*. US Army Laboratories, Natick, Massachusetts, Technical Report No. 70-21-ES, 78 pp.
Frost, D.V., 1967, Arsenicals in biology: retrospect and prospect. *Federation Proceedings*, Bethesda, Maryland, **26**(1), pp. 194–208.
Frye, R.H. & Wygant, N.D., 1971, Spruce beetle mortality in cacodylic acid-treated Engelmann spruce trap trees. *Journal of Economic Entomology*, **64**, pp. 911–16.
Fryer, J.D., 1974, *Effects of Herbicides in South Vietnam. B[10]. Uses of Herbicides in Tropics and Subtropics*. US National Academy of Sciences, Washington, 35 pp.
Fryer, J.D. & Evans, S.A., (eds.), 1968, *Weed Control Handbook. I. Principles. II. Recommendations*, 5th ed. Blackwell, Oxford. 2 vols. (494+325 pp.)
Fulbright, J.W., (ed.), 1970, *Background Information Relating to Southeast Asia and Vietnam*, 6th ed. US Senate, Washington, Committee on Foreign Relations, 455 pp.
Fulbright, J.W., (ed.), 1972, *Geneva Protocol of 1925*. US Senate, Washington, Committee on Foreign Relations, 439 pp.
Funk, D.T., 1962, *Revised Bibliography of Strip-mine Reclamation*. US Forest Service Central States Forest Experiment Station, Columbus, Ohio, Miscellaneous Release No. 35, 20 pp.
Futrell, R.F., Moseley, L.S. & Simpson, A.F., 1961, *United States Air Force in Korea 1950–1953*. Duell, Sloan & Pearce, New York, 774 pp. + pl.
Gaines, S.H., Vincent, F., Bloom, M. & Carter, J.F., 1938, *Bibliography on Soil Erosion and Soil Water Conservation: With Abstracts*. US Department of Agriculture, Washington, Miscellaneous Publication No. 312, 651 pp.
Garrett, W.E., 1974, No place to run: The Hmong of Laos. *National Geographic*, Washington, **145**, pp. 78–111.
Gee, J.G. & Anglemyer, M., 1970, *Vietnam Agriculture: A Selected Annotated Bibliography*. US Army Corps of Engineers, Washington, Engineer Agency for Resources Inventories, 58 pp.

Geiger, R., 1965, *Climate Near the Ground,* 4th ed. Harvard University Press, Cambridge, Massachusetts, 611 pp.
Gerking, S.D., 1948, Destruction of submerged aquatic plants by 2,4-D. *Journal of Wildlife Management,* Washington, **12**, pp. 221–27.
Glesinger, E., (ed.), 1962, *Forest Influences.* Food and Agriculture Organization of the United Nations, Rome, Forestry and Forest Products Studies No. 15, 307 pp.
Gliedman, J., 1972, *Terror From the Sky: North Viet-Nam's Dikes and the U.S. Bombing.* Vietnam Resource Center, Cambridge, Massachusetts, 172 pp.
Goldwater, B., (ed.), 1975, [Directives on the U.S. rules of engagement in Indochina.] *US Congressional Record,* Washington, **121**, pp. S9897–S9905.
Golley, F.B., McGinnis, J.T., Clements, R.G., Child, G.I. & Duever, M.J., 1969, Structure of tropical forests in Panama and Colombia. *BioScience,* Washington, **19**, pp. 693–96.
Golley, F.B., Odum, H.T. & Wilson, R.F., 1962, Structure and metabolism of a Puerto Rican red mangrove forest in May. *Ecology,* US, **43**, pp. 9–19.
Gómez-Pompa, A., Vázquez-Yanes, C. & Guevara, S., 1972, Tropical rain forest: a non-renewable resource. *Science,* Washington, **177**, pp. 762–65.
Goyer, G.G., Grant, L.O. & Henderson, T.J., 1966, Laboratory and field evaluation of Weathercord, a high output cloud seeding device. *Journal of Applied Meteorology,* Boston, **5**, pp. 211–16.
Gratkowski, H., 1974, *Herbicidal Drift Control: Aerial Spray Equipment, Formulations, and Supervision.* US Forest Service, Washington, General Technical Report No. PNW-14, 12 pp.
Gravel, M. *et al.,* (eds.), 1971–1972, *Pentagon Papers: The Defense Department History of United States Decisionmaking on Vietnam.* Beacon Press, Boston, 5 vols. (632+834+746+687+413 pp.)
Greenberg, D.S., 1972a, Defoliation: secret Army study urges use in future wars. *Science and Government Report,* Washington, **2**(11), pp. 1, 3–4.
Greenberg, D.S., 1972b, Vietnam rainmaking: a chronicle of DoD's snowjob. *Science and Government Report,* Washington, **2**(5), pp. 1, 4.
Gribble, G.W., 1974, TCDD: a deadly molecule. *Chemistry,* Washington, **47**(2), pp. 15–18.
Grimes, A.E., 1967, *Annotated Bibliography on the Climate of the Republic of Vietnam.* US Department of Commerce, Washington, Environmental Science Services Administration, Publication No. WB/BC-90, 122 pp.
Grimes, A.E., 1972, *Annotated Bibliography on Weather Modification 1960–1969.* US National Oceanic and Atmospheric Administration, Rockville, Maryland, Technical Memorandum No. EDS ESIC-1, 407 pp.
Grover, R., 1972, Effect of picloram on some soil microbial activities. *Weed Research,* Oxford, **12**, pp. 112–14.
Gupta, O.P., 1968, Herbicidal control in paddy fields. *World Crops,* London, **20**(5), pp. 25–27.

Hanson, W.R., 1952, Effects of some herbicides and insecticides on biota of North Dakota marshes. *Journal of Wildlife Management,* Washington, **16**, pp. 299–308.
Hare, R.C., 1961, *Heat Effects on Living Plants.* US Forest Service Southern Forest Experiment Station, New Orleans, Occasional Paper No. 183, 32 pp.
Harper, J.L., 1956–1957, Ecological aspects of weed control. *Outlook on Agriculture,* Bracknell, England, **1**, pp. 197–205.
Harrison, J.L., 1968, Effect of forest clearance on small mammals. In: Talbot, L.M. & Talbot, M.H., (eds.), *Conservation in Tropical South East Asia.* International Union for Conservation of Nature and Natural Resources, Morges, Switzerland, Publication New Series No. 10, 550 pp. + 2 pl., pp. 153–56.

Hartmann, K., 1967, [Chemical warfare 1966–67.] (In German) *Wehrkunde*, Munich, **16**, pp. 341–43.
Harvey, F., 1967, *Air War: Vietnam*. Bantam, New York, 186 pp. + 16 pl.
Haseltine, W. & Westing, A.H., 1971, The wasteland: beating plowshares into swords. *New Republic*, Washington, **165**(18), pp. 13–15.
Hassall, K.A., 1965, Pesticides: their properties, uses and disadvantages: general introduction; insecticides and related compounds; fungicides and herbicides; pesticides in relation to animals. *British Veterinary Journal*, London, **121**, pp. 105–18, 199–211.
Hawley, R.C. & Stickel, P.W., 1948, *Forest Protection*, 2nd ed. John Wiley, New York, 355 pp.
Hay, J.H., Jr., 1974, *Vietnam Studies: Tactical and Materiel Innovations*. US Department of the Army, Washington, 197 pp.
Headley, J.C. & Erickson, E., 1970, *The Pesticide Problem: An Annotated Bibliography*. University of Missouri Agricultural Experiment Station, Columbia, Missouri, Research Bulletin No. 970, 53 pp.
Heald, E.J. & Odum, W.E., 1970, Contribution of mangrove swamps to Florida fisheries. *Proceedings of the Gulf and Caribbean Fisheries Institute*, **22**, pp. 130–35.
Heath, R.G., Spann, J.W., Hill, E.F. & Kreitzer, J.F., 1972, *Comparative Dietary Toxicities of Pesticides to Birds*. US Bureau of Sport Fisheries and Wildlife, Washington, Special Scientific Report – Wildlife No. 152, 57 pp.
Heavner, R.O., 1970–1971, Interdiction: a dying mission? *Air University Review*, Maxwell Air Force Base, Alabama, **22**(2), pp. 56–59.
Helling, C.S., Isensee, A.R., Woolson, E.A., Ensor, P.D.J., Jones, G.E., Plimmer, J.R. & Kearney, P.C., 1973, Chlorodioxins in pesticides, soils, and plants. *Journal of Environmental Quality*, **2**, pp. 171–78.
Henderson, G.R.G., 1955, Whirling wings over the jungle. *Air Clues*, London, **9**, pp. 239–43.
Hersh, S.M., 1968, *Chemical and Biological Warfare: America's Hidden Arsenal*. Bobbs-Merrill, Indianapolis, 354 pp.
Hersh, S.M., 1972, Rainmaking is used as weapon by U.S. *New York Times*, 3 July 1972, pp. 1–2; 4 July 1972, p. 3.
Hewlett, J.D. & Nutter, W.L., 1969, *Outline of Forest Hydrology*. University of Georgia Press, Athens, Georgia, 137 pp.
Hibbert, A.R., 1969, Water yield changes after converting a forested catchment to grass. *Water Resources Research*, Washington, **5**, pp. 634–40.
Hilton, J.L., Bovey, R.W., Hull, H.M., Mullison, W.R. & Talbert, R.E., 1974, *Herbicide Handbook of the Weed Science Society of America*, 3rd ed. Weed Science Society of America, Champaign, Illinois, 430 pp.
Hobbs, C., 1964, *Southeast Asia: An Annotated Bibliography of Selected Reference Sources in Western Languages*, rev. ed. US Library of Congress, Washington, 180 pp.
Hobbs, C.C., Fuller, G.H., Jones, H.D., Dorosh, J.T. & Sacks, I.M., 1950, *Indochina: A Bibliography of the Land and People*. US Library of Congress, Washington, 367 pp.
Hobbs, P.V., Harrison, H. & Robinson, E., 1974, Atmospheric effects of pollutants. *Science*, Washington, **183**, pp. 909–15.
Hodges, H.S. & Webb, G.R., 1968, Climates of south-east Asia. *Australian Army Journal*, Canberra, **1968**(228), pp. 29–44.
Hodges, J.D. & Pickard, L.S., 1971, Lightning in the ecology of the southern pine beetle, *Dendroctonus frontalis* (Coleoptera: Scolytidae). *Canadian Entomologist*, Ottawa, **103**, pp. 44–51.
Hogg, O.F.G., 1970, *Artillery: Its Origin, Heyday and Decline*. C. Hurst, London, 330 pp. + ph.
Holden, A.V., 1964, Possible Effects on Fish of Chemicals Used in Agriculture. *Institute of Sewage Purification Journal and Proceedings*, London, 1964, pp. 361–68.

Holm, L. & Herberger, J., 1971, World list of useful publications for the study of weeds and their control. *Pans,* London, **17**(1), pp. 119–32.
Holmes, C.H., 1951, *Grass, Fern, and Savannah Lands of Ceylon, their Nature and Ecological Significance.* British Imperial Forestry Institute, Oxford, Paper No. 28, 95 pp. + 7 pl.
Hornbeck, J.W., Pierce, R.S. & Federer, C.A., 1970, Streamflow changes after forest clearing in New England. *Water Resources Research,* Washington, **6**, pp. 1124–32.
Horton, J.S., 1973, *Evapotranspiration and Watershed Research as Related to Riparian and Phreatophyte Management: An Abstract Bibliography.* US Department of Agriculture, Washington, Miscellaneous Publication No. 1234, 192 pp.
House, W.B., Goodson, L.H., Gadberry, H.M. & Dockter, K.W., 1967. *Assessment of Ecological Effects of Extensive or Repeated Use of Herbicides.* Midwest Research Institute, Kansas City, Missouri, 369 pp.
Howard, J.D., 1972, *Herbicides in support of counterinsurgency operations: a cost-effectiveness study.* US Naval Postgraduate School, Monterey, California, Master of Science thesis, 127 pp.
Hubbard, C.E., Whyte, R.O., Brown, D. & Gray, A.P., 1944, *Imperata cylindrica: Taxonomy, Distribution, Economic Significance and Control.* British Imperial Agricultural Bureaux, Aberystwyth, Joint Publication No. 7, 63 pp.
Huff, J.E. & Wassom, J.S., 1973, Chlorinated dibenzodioxins and dibenzofurans. *Environmental Health Perspectives,* Research Triangle Park, North Carolina, **1973**(5), pp. 283–312.
Hughes, J.F., 1949–1950, Influence of forests on climate and water supply. I. Forests and climate. II. Forests and water supply. *Forestry Abstracts,* Farnham Royal, England, **11,** pp. 145–53, 283–92.
Hull, A.C., Jr., 1971, Effect of spraying with 2,4-D upon abundance of pocket gophers in Franklin Basin, Idaho. *Journal of Range Management,* Denver, **24**, pp. 230–32.
Hunter, J.H. & Young, A.L., 1972, *Vegetative succession studies on a defoliant-equipment test area, Eglin AFB Reservation, Florida.* US Air Force Armament Laboratory, Eglin Air Force Base, Florida, Technical Report No. 72–31, 23 pp.
Hursh, C.R., 1948, *Local Climate in the Copper Basin of Tennessee as Modified by the Removal of Vegetation.* US Department of Agriculture, Washington, Circular No. 774, 38 pp.
Hutnik, R.J. & Davis, G., (eds.), 1973, *Ecology and Reclamation of Devastated Land.* Gordon and Breach, New York, 2 vols. (538+504 pp.)
Hymoff, E., 1971, Technology vs. guerrillas: stalemate in Indo-china. *Bulletin of the Atomic Scientists,* Chicago, **27**(9), pp. 27–30.
Ingram, W.M. & Tarzwell, C.M., 1954, *Selected Bibliography of Publications Relating to Undesirable Effects Upon Aquatic Life by Algicides, Insecticides, Weedicides.* US Public Health Service, Washington, Publication No. 400, 28 pp.
Irish, K.R., Darrow, R.A. & Minarik, C.E., 1969, *Information Manual for Vegetation Control in Southeast Asia.* US Army Fort Detrick, Frederick, Maryland, Miscellaneous Publication No. 33, 71 pp.
Ise, J., 1920, *United States Forest Policy.* Yale University Press, New Haven, 395 pp.
Isensee, A.R. & Jones, G.E., 1971, Absorption and translocation of root and foliage applied 2,4-dichlorophenol, 2,7-dichlorodibenzo-*p*-dioxin, and 2,3,7,8-tetrachlorodibenzo-*p*-dioxin. *Journal of Agricultural and Food Chemistry,* Washington, **19**, pp. 1210–14.
Isensee, A.R., Kearney, P.C., Woolson, E.A., Jones, G.E. & Williams, V.P., 1973, Distribution of alkyl arsenicals in model ecosystem. *Environmental Science and Technology,* Washington, **7**, pp. 841–45.
Israelyan, V., 1974, New Soviet initiative on disarmament. *International Affairs,* Moscow, **1974**(11), pp. 19–25.

Jackson, M.L., Gillette, D.A., Danielsen, E.F., Blifford, I.H., Bryson, R.A. & Syers, J.K., 1973, Global dustfall during the quaternary as related to environments. *Soil Science,* Baltimore, **116**, pp. 135–45.

Janzen, D.H., 1973, Tropical agroecosystems. *Science,* Washington, **182**, pp. 1212–19.

Jenny, H., Gessel, S.P. & Bingham, F.T., 1949, Comparative study of decomposition rates of organic matter in temperate and tropical regions. *Soil Science,* Baltimore, **68**, pp. 419–32.

Johnson, L.R. & Hiltbold, A.E., 1969, Arsenic content of soil and crops following use of methanearsonate herbicides. *Soil Science Society of America Proceedings,* Madison, Wisconsin, **33**, pp. 279–82.

Johnston, D.W. & Odum, E.P., 1956, Breeding bird populations in relation to plant succession on the piedmont of Georgia. *Ecology,* US, **37**, pp. 50–62.

Johnstone, L.C., 1970–1971, Ecocide and the Geneva Protocol. *Foreign Affairs,* New York, **49**, pp. 711–20.

Jones, G.R.N., 1972, CS and its chemical relatives. *Nature,* London, **235**, pp. 257–61.

Jordan, C.F. & Kline, J.R., 1972, Mineral cycling: some basic concepts and their application in a tropical rain forest. *Annual Reviews of Ecology and Systematics,* **3**, pp. 33–50.

Jordan, C.F., Kline, J.R. & Sasscer, D.S., 1972, Relative stability of mineral cycles in forest ecosystems. *American Naturalist,* Chicago, **106**, pp. 237–53.

Juntunen, E.T. & Norris, L.A., 1972, *Field Application of Herbicides: Avoiding Danger to Fish.* Oregon State University Agricultural Experiment Station, Corvallis, Oregon, Special Report No. 354, 26 pp.

Kamm, H., 1971, U.S. pilots find enemy traffic moving freely on trail in Laos. *New York Times,* 11 April 1971, pp. 1, 4.

Kasasian, L., 1971, *Weed Control in the Tropics.* Leonard Hill, London, 307 pp. + 51 pl.

Katz, Y.H., 1966, *Nuclear War and Soil Erosion: Some Problems and Prospects.* Rand Corporation, Santa Monica, California, Memorandum No. RM-5203-TAB, 87 pp.

Kearney, P.C. & Helling, C.S., 1969, Reactions of pesticides in soils. *Residues Reviews,* **25**, pp. 25–44.

Kearney, P.C., Isensee, A.R., Helling, C.S., Woolson, E.A. & Plimmer, J.R., 1973a, Environmental significance of chlorodioxins. In: Blair, E.H., (ed.), *Chlorodioxins: Origin and Fate.* American Chemical Society, Washington, Advances in Chemistry Series No. 120, 141 pp., pp. 105–11.

Kearney, P.C., Nash, R.G. & Isensee, A.R., 1969a, Persistence of pesticide residues in soils. In: Miller, M.W. & Berg, G.G., (eds.), *Chemical Fallout: Current Research on Persistent Pesticides.* C.C. Thomas, Springfield, Illinois, 532 pp., pp. 54–67.

Kearney, P.C., Woolson, E.A. & Ellington, C.P., Jr., 1972, Persistence and metabolism of chlorodioxins in soils. *Environmental Science and Technology,* Washington, **6**, pp. 1017–19.

Kearney, P.C., Woolson, E.A., Isensee, A.R. & Helling, C.S., 1973b, Tetrachlorodibenzodioxin in the environment: sources, fate, and decontamination. *Environmental Health Perspectives,* Research Triangle Park, North Carolina, **1973**(5), pp. 273–77.

Kearney, P.C., Woolson, E.A., Plimmer, J.R. & Isensee, A.R., 1969b, Decontamination of pesticides in soils. *Residue Reviews,* **29**, pp. 137–49.

Keith, J.O., Hansen, R.M. & Ward, A.L., 1959, Effect of 2,4-D on abundance and foods of pocket gophers. *Journal of Wildlife Management,* Washington, **23**, pp. 137–45.

Kennedy, E.M., (ed.), 1967, *Civilian Casualty, Social Welfare, and Refugee Problems in South Vietnam.* US Senate, Washington, Committee on the Judiciary, 332 pp.

Kennedy, E.M., (ed.), 1971, *War-related Civilian Problems in Indochina. I. Vietnam. II. Laos and Cambodia. III. Vietnam.* US Senate, Washington, Committee on the Judiciary, 154+173+91 pp.

Kerver, T.J., 1974, To clear the way. *National Defense,* Washington, **58**, pp. 454–55.
Kieffer, F.V., 1972, *Bibliography of Surface Coal Mining in the United States to August, 1971.* Forum Associates, Columbus, Ohio, 71 pp.
Kiernan, J.M., 1967, Combat engineers in the Iron Triangle. *Army,* Washington, **17**(6), pp. 42–45.
Kimbrough, R.D., 1972, Toxicity of chlorinated hydrocarbons and related compounds: a review including chlorinated dibenzodioxins and chlorinated dibenzofurans. *Archives of Environmental Health,* Chicago, **25**, pp. 125–31.
Kimmins, J.P. & Fraker, P.N., 1973, *Bibliography of Herbicides in Forest Ecosystems.* Canadian Forestry Service, Ottawa, Information Report No. BC-X-81, 261 pp.
Kipp, D.H., 1931, Wild life in a fire. *American Forests,* Washington, **37**, pp. 323–25, 360.
Kipp, R.M., 1967–1968, Counterinsurgency from 30,000 feet: the B-52 in Vietnam. *Air University Review,* Maxwell Air Force Base, Alabama, **19**(2), pp. 10–18.
Kira, T. & Shidei, T., 1967, Primary production and turnover of organic matter in different forest ecosystems of the western Pacific. *Japanese Journal of Ecology,* Sendai, **17**, pp. 70–87.
Kittredge, J., 1948, *Forest Influences: The Effects of Woody Vegetation on Climate, Water, and Soil, with Applications to the Conservation of Water and the Control of Floods and Erosion.* McGraw-Hill, New York, 394 pp.
Klingman, D.L. & Shaw, W.C., 1971, *Using Phenoxy Herbicides Effectively: 2,4-D, 2,4,5-T, MCPA, Silvex, 2,4-DB,* rev.ed. US Department of Agriculture, Washington, Farmers' Bulletin No. 2183, 24 pp.
Kopp, M.L., 1972, [Letter on war damage to South Vietnam.]. *US Congressional Record,* Washington, **118**, pp. 12241–42.
Kotsch, W.J., 1968, Forecast: change. *US Naval Institute Proceedings,* Annapolis, Maryland, **94**(1), pp. 69–77.
Kozlowski, T.T. & Ahlgren, C.E., (eds.), 1974, *Fire and Ecosystems.* Academic Press, New York, 542 pp.
Kristoferson, L., (ed.), 1975, War and environment: a special issue. *Ambio,* Stockholm, **4**(5–6), pp. 178–244.
Kuenen, D.J., (ed.), 1961, *Ecological Effects of Biological and Chemical Control of Undesirable Plants and Animals.* E.J. Brill, Leiden, 118 pp. + 3 pl.
Kuronuma, K., 1961, *Check List of Fishes of Vietnam.* US Operations Mission Division of Agriculture and Natural Resources, Saigon, 66 pp.
Kutzleb, C.R., 1971–1972, Can forests bring rain to the plains? *Forest History,* Santa Cruz, California, **15**(3), pp. 14–21.
Lacoste, Y., 1972, Bombing the dikes: a geographer's on-the-site analysis. *Nation,* New York, **215**, pp. 298–301.
Landsberg, H.E., 1970, Man-made climatic changes. *Science,* Washington, **170**, pp. 1265–74.
Lang, A. *et al.,* 1974, *Effects of Herbicides in South Vietnam. A. Summary and conclusions.* US National Academy of Sciences, Washington, [398] pp. + 8 maps.
Laurie, M.V., 1974, *Tree Planting Practices in African Savannas.* Food and Agriculture Organization of the United Nations, Rome, Forestry Development Paper No. 19, 185 pp. + 3 maps.
Lawrence, G.H.M., 1951, *Taxonomy of Vascular Plants.* Macmillan, New York, 823 pp.
Lawrence, J.M. & Hollingsworth, E.B., 1962–1969, *Aquatic Herbicide Data.* US Department of Agriculture, Washington, Agriculture Handbook No. 231, 133+126 pp.
Lehn, P.J., Young, A.L., Hamme, N.A. & Wolverton, B.C., 1970, *Studies to Determine the Presence of Artificially Induced Arsenic Levels in Three Freshwater Streams and its Effect on Fish Species Diversity.* US Air Force Armament Laboratory, Eglin Air Force Base, Florida, Technical Report No. 70–81, 27 pp.

Leitenberg, M. & Burns, R.D., 1973, *Vietnam Conflict: its Geographical Dimensions, Political Traumas, & Military Developments.* ABC Clio, Santa Barbara, California, 164 pp.

Leopold, A., 1933, *Game Management.* Charles Scribner's Sons, New York, 481 pp. + 3 pl.

Leopold, A.S., 1950, Deer in relation to plant successions. *Journal of Forestry,* Washington, **48**, pp. 675–78.

Leung, W.-T.W., Butrum, R.R. & Chang, F.H., 1972, *Selected Bibliography on East-Asian Foods and Nutrition arranged according to Subject Matter and Area.* US Department of Health, Education, and Welfare, Washington, Publication No. (NIH)73-466, 296 pp.

Lewallen, J., 1971, *Ecology of Devastation: Indochina.* Penguin, Baltimore, 179 pp.

Library of Congress, US, 1971, *Impact of the Vietnam War.* US Senate, Washington, Committee on Foreign Relations, 36 pp.

Life Magazine, 1964, First World War. VI. Scarred face of Verdun. *Life Magazine,* Chicago, **56**(23), pp. 68–83.

Life Magazine, 1971, Bomb called 'cheeseburger'. *Life Magazine,* Chicago, **70**(19), pp. 40–41.

Likens, G.E. & Bormann, F.H., 1974, Linkages between terrestrial and aquatic ecosystems. *BioScience,* Washington, **24**, pp. 447–56.

Likens, G.E., Bormann, F.H., Johnson, N.M., Fisher, D.W. & Pierce, R.S., 1970, Effects of forest cutting and herbicide treatment on nutrient budgets in the Hubbard Brook watershed-ecosystem. *Ecological Monographs,* US, **40**, pp. 23–47.

Littauer, R. & Uphoff, N., (eds.), 1972, *Air War in Indochina,* rev. ed. Beacon Press, Boston, 289 pp.

Little, E.C.S. & Ivens, G.W., 1965, Control of brush by herbicides in tropical and subtropical grassland. I. Americas, Australia, the Pacific, New Zealand. II. Africa. *Herbage Abstracts,* Farnham Royal, England, **35**(1), pp. 1–12.

Liu, L.C. & Cibes-Viade, H.R., 1972, Effect of various herbicides on the respiration of soil microorganisms. *Journal of Agriculture of the University of Puerto Rico,* Rio Piedras, **56**, pp. 417–25.

Livingstone, D.A. *et al.,* 1966, Biological aspects of weather modification. *Bulletin of the Ecological Society of America,* **47**, pp. 39–78.

Lofroth, G., 1970, *Alvsbyn Reindeer Mortality.* Smithsonian Institution Center for Short-lived Phenomena, Cambridge, Massachusetts, Event No. 51–70, 1 p.

Loftas, T., 1970, Fishery Boom for South Vietnam? *New Scientist,* London, **46**, pp. 280–83.

Lohs, Kh., 1973a, [Dioxin: a new warfare agent of imperialist armies?] (In German) *Zeitschrift für Militärmedizin, East Berlin,* **14**, pp. 318–19.

Lohs, Kh., 1973b, Fire as a means of warfare. *Scientific World,* London, **17**(1), p. 18.

Loos, M.A., 1969, Phenoxyalkanoic acids. In: Kearney, P.C. & Kaufman, D.D., (eds.), *Degradation of Herbicides.* Marcel Dekker, New York, 394 pp., pp. 1–49.

Lowdermilk, W.C., 1930, Influence of forest litter on run-off, percolation, and erosion. *Journal of Forestry,* Washington, **28**, pp. 474–91.

Lowry, W.P., 1967, Climate of cities. *Scientific American,* New York, **217**(2), pp. 15–23, 128.

Lucas, E.H. & Hamner, C.L., 1947, Inactivation of 2,4-D by adsorption on charcoal. *Science,* Washington, **105**, p. 340.

Lugo, A.E. & Snedaker, S.C., 1974, Ecology of mangroves. *Annual Reviews of Ecology and Systematics,* **5**, pp. 39–64.

Lull, H.W. & Reinhart, K.G., 1972, *Forests and Floods in the Eastern United States.* US Forest Service, Washington, Research Paper No. NE-226, 94 pp.

Lutz, H.J., 1956, *Ecological Effects of Forest Fires in the Interior of Alaska.* US Depart-

ment of Agriculture, Washington, Technical Bulletin No. 1133, 121 pp.
Lutz, H.J. & Chandler, R.F., Jr., 1946. *Forest Soils.* John Wiley, New York, 514 pp.
Lutz-Ostertag, Y. & Lutz, H., 1970, [Deleterious action of the herbicide 2,4-D on the embryonic development and fecundity of game birds.] (In French) *Comptes Rendus Hebdomadaires des Séances, Académie des Sciences,* Paris, **271**(D), pp. 2418–21 + 2 pl.
Lynn, G.E., 1964–1965, Review of toxicological information on tordon herbicides. *Down to Earth,* Midland, Michigan, **20**(4), pp. 6–8.
Lyons, A.R., Coblentz, T.H. & Waterstradt, W., 1971, *Effectiveness of Defoliation Operations for Temperate Zone Vegetation.* US Army Fort Douglas, Utah, Desert Test Center Technical Report No. 71–116.
Machta, L. & Telegadas, K., 1974, Inadvertent large-scale weather modification. In: Hess, W.N., (ed.), *Weather and Climate Modification.* John Wiley, New York, 842 pp., pp. 687–725.
Macnae, W., 1968, General account of the fauna and flora of mangrove swamps and forests in the Indo-West-Pacific region. *Advances in Marine Biology,* **6**, pp. 73–270 + 1 fig.
Madge, D.S., 1965, Leaf fall and litter disappearance in a tropical forest. *Pedobiologia,* Jena, **5**, pp. 273–88.
Madgwick, H.A.I. & Ovington, J.D., 1959, Chemical composition of precipitation in adjacent forest and open plots. *Forestry,* London, **32**, pp. 14–22.
Mahon, G.H., (ed.), 1969, *Department of Defense Appropriations for 1970.* US House of Representatives, Washington, Committee on Appropriations, Part 6, 939 pp.
Mahony, F.J., 1975, *Draft Protocol I, Part IV, Section I, Article 48 bis.* Diplomatic Conference on the Reaffirmation and Development of International Humanitarian Law Applicable in Armed Conflicts, Geneva, Document No. CDDH/III/GT/35, 6 pp.
Maignien, R., 1966, *Review of Research on Laterites.* United Nations Scientific, Educational and Cultural Organization, Paris, Natural Resources Research Series No. 4, 148 pp. + 8 pl.
Mammen, F.v., 1916, *[Significance of the Forest Particularly in War.]* (In German) "Globus", Wissenschaftliche Verlagsanstalt, Dresden, 96 pp.
Manucy, A., 1949, *Artillery Through the Ages: A short Illustrated History of Cannon, Emphasizing Types Used in America.* US National Park Service, Washington, Interpretive Series, History, No. 3, 92 pp.
Marine Corps, US, 1974, *Marines in Vietnam 1954–1973: An Anthology and Annotated Bibliography.* US Marine Corps, Washington, History and Museums Division, 277 pp.
Marks, P.L. & Bormann, F.H., 1972, Revegetation following forest cutting: mechanisms for return to steady-state nutrient cycling. *Science,* Washington, **176**, pp. 914–15.
Marsh, G.P., 1864, *Man and Nature.* Harvard University Press, Cambridge, Massachusetts, 472 pp., 1965.
Massey, D., 1970, Land clearing team: Rome plows on the border. *Hurricane,* Saigon, **1970**(35), pp. 34–37.
Matheson, M.H. & Laurendeau, P.H., 1953, *Indo-China: A Geographical Appreciation.* Canada Department of Mines and Technical Surveys, Ottawa, Foreign Geography Information Series No. 6, 88 pp. + 26 figs.
Mattingly, R. & McVann, D., 1971, Back to Batangan. *Kysu',* Saigon, **3**(1), pp. 2–6.
McBride, B.C. & Wolfe, R.S., 1971, Biosynthesis of dimethylarsine by methanobacterium. *Biochemistry,* Washington, **10**, pp. 4312–17.
McCarthy, R.D., 1969, Ban on gas and germ warfare. *US Congressional Record,* Washington, **115**, pp. 15763–66.
McColl, J.G. & Grigal, D.F., 1975, Forest fire: effects on phosphorus movement to

lakes. *Science,* Washington, **188**, pp. 1109–11.
McConnell, A.F., Jr., 1969–1970, Mission: Ranch Hand. *Air University Review,* Maxwell Air Force Base, Alabama, **21**(2), pp. 89–94.
McDonald, J.E., 1962, Evaporation-precipitation fallacy. *Weather,* London, **17**, pp. 168–77, 216.
McKinley, T.W., 1957, *Forests of Free Viet-Nam: A Preliminary Study for Organization, Protection, Policy and Production.* US Operations Mission, Saigon, 153 pp.
McNeil, M., 1964, Lateritic soils. *Scientific American,* New York, **211**(5), pp. 96–102, 154, 156.
Meijer, W., 1970, Regeneration of tropical lowland forest in Sabah, Malaysia forty years after logging. *Malayan Forester,* Kepong, Salangor, **33**, pp. 204–29.
Meijer, W., 1973, Devastation and regeneration of lowland dipterocarp forests in southeast Asia. *BioScience,* Washington, **23**, pp. 528–33.
Meikle, R.W., Youngson, C.R., Hedlund, R.T., Goring, C.A.I., Hamaker, J.W. & Addington, W.W., 1973, Measurement and prediction of picloram disappearance rates from soil. *Weed Science,* Urbana, Illinois, **21**, pp. 549–55.
Mekong Documentation Centre of the Committee for Coordination of Investigations of the Lower Mekong Basin, 1967, *Cambodia: A Selected Bibliography.* United Nations Economic Commission for Asia and the Far East, Bangkok, Publication No. WRD/MKG/INF/L.211, 101 pp.
Mellanby, K., 1970, *Pesticides and Pollution,* 2nd ed. Collins, London, 221 pp. + 14 pl.
Meselson, M.S., Westing, A.H., Constable, J.D. & Cook, R.E., 1971, Preliminary report of Herbicide Assessment Commission of the American Association for the Advancement of Science. In: Kennedy, E.M., (ed.), *War-related Civilian Problems in Indochina. I. Vietnam.* US Senate, Washington, Committee on the Judiciary, 154 pp., pp. 113–31.
Milinkó, I., 1967, [Contribution to the problem of virus diseases of *Capsicum grossum.*] (In Russian) In: Kapeller, K., (ed.), *[Fiftieth Scientific Jubilee Session on* Capsicum grossum.] Magy. Agrártud. Egyes. Duna-Tisza Köz. Mezög. Kísér. Inér., Kalocsa, Hungary, 274 pp., pp. 131–35 + 1 tbl. *Cf. Review of Applied Mycology,* Kew, England, **48**, Abstract No. 2689c, 1969.
Minarik, C.E. & Bertram, A.L., 1962, *Evaluation of the Ca Mau Peninsula Defoliation Targets 9 November 1962 in Republic of Vietnam.* US Army Fort Detrick, Frederick, Maryland, Biological Laboratories, 5 pp.
Minarik, C.E. & Darrow, R.A., 1968, Persistence of herbicides in soil and water. In: Herbicide Policy Review Committee (ed.), *Report on the Herbicide Policy Review.* US Embassy, Saigon, 80 pp. + 5 apps. (12+13+5+12+6 pp.), Appendix D, 12 pp.
Misra, S.G. & Tiwari, R.C., 1963, Studies on arsenite-arsenate adsorption on soils. *Indian Journal of Applied Chemistry,* Calcutta, **26**, pp. 117–21.
Mizuno, T. & Mori, S., 1970, Preliminary hydrobiological survey of some southeast Asian inland waters. *Biological Journal of the Linnean Society,* London, **2**, pp. 77–118 + 4 pl.
Mobley, H.E., 1974, Fire: its impact on the environment. *Journal of Forestry,* Washington, **72**, pp. 414–17.
Moffett, J.O., Morton, H.L. & Macdonald, R.H., 1972, Toxicity of some herbicidal sprays to honey bees. *Journal of Economic Entomology,* U.S., **65**, pp. 32–36.
Mohr, E.C.J., Baren, F.A.v. & Schuylenborgh, J.v., 1972, *Tropical Soils: A Comprehensive Study of their Genesis,* 3rd ed. Mouton, Ichtiar Baru, van Hoeve, The Hague, 481 pp. + pl.
Mokma, D.L., Syers, J.K., Jackson, M.L., Clayton, R.N. & Rex, R.W., 1972, Aeolian additions to soils and sediments in the south Pacific area. *Journal of Soil Science,* UK, **23**, pp. 147–62 + 1 pl.
Moore, J.A., (ed.), 1973, Perspective on chlorinated dibenzodioxins and dibenzofurans.

Environmental Health Perspectives, Research Triangle Park, North Carolina, **1973**(5), 313 pp.
Moore, N.W., (ed.), 1966, Pesticides in the environment and their effects on wildlife. *Journal of Applied Ecology*, Oxford, **3**(Supplement), 311 pp.
Moore, N.W., 1967, Synopsis of the pesticide problem. *Advances in Ecological Research*, **4**, pp. 75–129.
Moormann, F.R., 1961, *Soils of the Republic of Viet-Nam*. Republic of Vietnam Ministry of Agriculture, Saigon, 66 pp. + 1 map.
Morton, H.L., Moffett, J.O. & Macdonald, R.H., 1972, Toxicity of herbicides to newly emerged honey bees. *Environmental Entomology*, **1**(1), pp. 102–104.
Motooka, P.S., Saiki, D.F., Plucknett, D.L., Younge, O.R. & Daehler, R.E., 1967, *Aerial Herbicidal Control of Hawaii Jungle Vegetation*. Hawaii Agricultural Experiment Station, Honolulu, Bulletin No. 140, 19 pp.
Mrak, E.M. *et al.*, 1969–1972, *Report of the Secretary's Commission on Pesticides and their Relationship to Environmental Health*. US Department of Health, Education, and Welfare, Washington, 2 vols. (677+310 pp.)
Mullison, W.R., 1970, Effects of herbicides on water and its inhabitants. *Weed Science*, Urbana, Illinois, **18**, pp. 738–50.
Mutch, R.W., 1970, Wildland fires and ecosystems: a hypothesis. *Ecology*, U.S., **51**, pp. 1046–51.
Myrdal, A. *et al.*, 1972, *Disarmament and Development*. United Nations, New York, 37 pp.
National Academy of Sciences, 1974, *Effects of Herbicides in South Vietnam*. US National Academy of Sciences, Washington, 20 vols. ([1841] pp.)
National Defense, 1973–1974, Bronco and the FAE. *National Defense*, Washington, **58**, p. 174.
National Geographic Service, 1969, *Republic of Vietnam Vegetation Map*. Republic of Vietnam National Geographic Service, Dalat, map, 73 x 105 cm.
National Geographic Society, 1967, *Vietnam, Cambodia, Laos, and Thailand*. National Geographic Society, Washington, map, 80 x 100 cm + index, 35 pp.
Neilands, J.B., 1970a, Ecocide in Vietnam. In: Brown, S. & Acklund, L., (eds.), *Why Are We Still in Vietnam?* Vintage, New York, 144 pp., pp. 87–97.
Neilands, J.B., 1970b, Vietnam: progress of the chemical war. *Asian Survey*, Berkeley, California, **10**, pp. 209–29.
Neilands, J.B., 1972a, Gas warfare in Vietnam in perspective. In: Neilands, J.B. *et al.*, *Harvest of Death: Chemical Warfare in Vietnam and Cambodia*. Free Press, New York, 304 pp., pp. 3–101.
Neilands, J.B., 1972b, Napalm survey. In: Browning, F. & Forman, D., (eds.), *Wasted Nations*. Harper and Row, New York, 346 pp., pp. 26–37.
Neilands, J.B., 1973, Survey of chemical and related weapons. *Naturwissenschaften*, West Berlin, **60**, pp. 177–83.
Neubert, D., Zens, P., Rothenwallner, A. & Merker, H.-J., 1973, Survey of the embryotoxic effects of TCDD in mammalian species. *Environmental Health Perspectives*, Research Triangle Park, North Carolina, **1973**(5), pp. 67–79.
Newell, R.E., 1971, Amazon forest and atmospheric general circulation. In: Matthews, W.H., Kellogg, W.W. & Robinson, G.D., (eds.), *Man's Impact on the Climate*. MIT Press, Cambridge, Massachusetts, 594 pp., pp. 457–59.
Newman, A.S., Thomas, J.R. & Walker, R.L., 1952, Disappearance of 2,4-dichlorophenoxyacetic acid and 2,4,5-trichlorophenoxyacetic acid from soil. *Soil Science Society of American Proceedings*, Madison, Wisconsin, **16**, pp. 21–24.
Ngan, Phung Trung., 1968, Status of conservation in South Vietnam. In: Talbot, L.M. & Talbot, M.H., (eds.), *Conservation in Tropical South East Asia*. International Union for Conservation of Nature and Natural Resources, Morges, Switzerland, Publication

New Series No. 10, 550 pp. + 2 pl., pp. 519–22.
Nihart, B., 1970–1971, Ranch Hand may ride again: opium poppies the target? *Armed Forces Journal,* Washington, **108**(21), pp. 19–20.
Norris, L.A., 1970, Degradation of herbicides in the forest floor. In: Youngberg, C.T. & Davey, C.B., (eds.), *Tree Growth and Forest Soils.* Oregon State University Press, Corvallis, Oregon, 544 pp., pp. 397–411.
Norris, L.A. & Miller, R.A., 1974, Toxicity of 2,3,7,8-tetrachlorodibenzo-*p*-dioxin (TCDD) in guppies (*Poecilia reticulatus* Peters). *Bulletin of Environmental Contamination and Toxicology,* West Berlin, **12**, pp. 76–80.
Novy, D.S. & Majumdar, S.K., 1972, Effects of 2,4,5-trichlorophenoxyacetic acid on *Drosophila melanogaster. Proceedings of the Pennsylvania Academy of Science,* **46**, pp. 21–22.
Nuttonson, M.Y., 1963, *Climatological Data of Vietnam, Laos and Cambodia.* American Institute of Crop Ecology, Silver Spring, Maryland, Publication No. 29C, 75 pp.
Nye, P.H., 1961, Organic matter and nutrient cycles under moist tropical forest. *Plant and Soil,* The Hague, **13**, pp. 333–46.
Nye, P.H. & Greenland, D.J., 1960, *Soil Under Shifting Cultivation.* British Commonwealth Bureau of Soils, Harpenden, Technical Communication No. 51, 156 pp. + 16 pl.
Nye, P.H. & Greenland, D.J., 1964, Changes in the soil after clearing tropical forest. *Plant and Soil,* The Hague, **21**, pp. 101–12.

Oakes, A.J., Jr., 1967, *Some harmful plants of southeast Asia.* US National Naval Medical Center, Bethesda, Maryland, 50 pp.
Odum, E.P., 1971, *Fundamentals of Ecology,* 3rd ed. W.B. Saunders, Philadelphia, 574 pp.
Odum, H.T., Sell, M., Brown, M., Zucchetto, J., Swallows, C., Browder, J., Ahlstrom, T. & Peterson, L., 1974, *Effects of Herbicides in South Vietnam. B[13]. Models of Herbicide, Mangroves, and War in Vietnam.* US National Academy of Sciences, Washington, 302 pp.
Ohman, H.L., 1965, *Climatic Atlas of Southeast Asia.* US Army Laboratories, Natick, Massachusetts, Technical Report No. ES-19, 7 pp. + 87 maps.
Oliver, K.H., Jr., Parsons, G.H. & Huffstetler, C.T., 1966, *Ecological Study on the Effects of Certain Concentrations of Cacodylic Acid on Selected Fauna and Flora.* Vitro Services of America, Eglin Air Force Base, Florida, Report No. APGC-TR-66-54, 23 pp.
Ordnance, 1971–1972, Jungle flatteners. *Ordnance,* Washington, **56**, p. 157.
Orians, G.H. & Pfeiffer, E.W., 1970, Ecological effects of the war in Vietnam. *Science,* Washington, **168**, pp. 544–54.
Orsi, J.J., 1974, Check list of the marine and freshwater fishes of Vietnam. *Publications of the Seto Marine Biological Laboratory,* **31**, pp. 153–77.
Orth, H., 1954, [On the inactivation of growth-regulating herbicides via adsorption onto charcoal.] (In German) *Zeitschrift für Pflanzenkrankheiten und Pflanzenschutz,* Stuttgart, **61**, pp. 385–96.
Otterman, J., 1974, Baring high-albedo soils by overgrazing: a hypothesized desertification mechanism. *Science,* Washington, **186**, pp. 531–33. *Cf. ibid.* **189**, pp. 1012–15.

Page, C.H. & Vigoureux, P., (eds.), 1972, *International System of Units (SI).* US National Bureau of Standards, Washington, Special Publication No. 330, 42 pp.
Paine, R.T., 1966, Food web complexity and species diversity. *American Naturalist,* Chicago, **100**, pp. 65–75.
Palm, C.E. *et al.*, 1968, *Weed Control.* US National Academy of Sciences, Washington, Publication No. 1597, 471 pp.
Palmer, J.S., 1972, *Toxicity of 45 Organic Herbicides to Cattle, Sheep, and Chickens.* US

Agricultural Research Service, Beltsville, Maryland, Production Research Report No. 137, 41 pp.
Palmer, J.S. & Radeleff, R.D., 1969, *Toxicity of Some Organic Herbicides to Cattle, Sheep, and Chickens.* US Agricultural Research Service, Beltsville, Maryland, Production Research Report No. 106, 26 pp.
Palmer-Jones, T., 1964, Effect on honey bees of 2,4-D. *New Zealand Journal of Agricultural Research,* Wellington, **7**, pp. 339–42.
Pate, B.D., Voigt, R.C., Lehn, P.J. & Hunter, J.H., 1972, *Animal Survey Studies of Test Area C-52A, Eglin AFB Reservation, Florida.* US Air Force Armament Laboratory, Eglin Air Force Base, Florida, Technical Report No. 72–72, 13 pp.
Patric, J.H., 1973, *Deforestation Effects on Soil Moisture, Streamflow, and Water Balance in the Central Appalachians.* US Forest Service, Washington, Research Paper No. NE-259, 12 pp.
Patric, J.H., 1974, River flow increases in central New England after the hurricane of 1938. *Journal of Forestry,* Washington, **72**, pp. 21–25.
Pell, C., (ed.), 1972a, *Prohibiting Military Weather Modification.* US Senate, Washington, Committee on Foreign Relations, 162 pp.
Pell, C., 1972b, Senate Resolution No. 281, 92nd Congress. *US Congressional Record,* Washington, **118**, pp. 8871–8873, 9277, 13386.
Pell, C., 1973a, *Prohibiting Environmental Modification as a Weapon of War.* US Senate, Washington, Report No. 93–270, 7 pp.
Pell, C., 1973b, Senate Resolution No. 71, 93rd Congress. *US Congressional Record,* Washington, **119**, pp. 23303–23305.
Pell, C., (ed.), 1974, *Weather Modification.* US Senate, Washington, Committee on Foreign Relations, 123 pp.
Pendleton, R.L., 1940, Soil erosion in the tropics. *Journal of Forestry,* Washington, **38**, pp. 753–62.
Pendleton, R.L., 1941, Laterite and its structural uses in Thailand and Cambodia. *Geographical Review,* New York, **31**, pp. 177–202.
Pendleton, R.L., 1953–1956, Place of tropical soils in feeding the world. *Ceiba,* **4**, pp. 201–22.
Penman, H.L., 1963, *Vegetation and Hydrology.* British Commonwealth Bureau of Soils, Harpenden, Technical Communication No. 53, 124 pp.
Perry, T.O., 1968, Vietnam: truths of defoliation. *Science,* Washington, **160**, p. 601.
Peterson, G.E., 1967, Discovery and development of 2,4-D. *Agricultural History,* Davis, California, **41**, pp. 243–53.
Pfeiffer, E.W., 1969–1970, Defoliation and bombing effects in Vietnam. *Biological Conservation,* Barking, England, **2**, pp. 149–51.
Pfeiffer, E.W., 1971, Craters. *Environment,* Saint Louis, **13**(9), pp. 3–8.
Pfeiffer, E.W., 1972, *Ecocide: A Strategy of War.* Thorne Films, Boulder, Colorado, film, 16 mm, colour, sound, 21 min.
Pfeiffer, E.W. & Orians, G.H., 1972, Military uses of herbicides in Vietnam. In: Neilands, J.B. *et al., Harvest of Death: Chemical Warfare in Vietnam and Cambodia.* Free Press, 304 pp., pp. 117–76.
Philpot, C.W. & Mutch, R.W., 1968, *Flammability of Herbicide-treated Guava Foliage.* US Forest Service, Washington, Research Paper No. INT-54, 9 pp.
Pimental, D., 1971, *Ecological Effects of Pesticides on Non-target Species.* US Office of Science and Technology, Washington, 220 pp.
Ploger, R.R., 1968, Different war: same old ingenuity. *Army,* Washington, **18**(9), pp. 70–75.
Ploger, R.R., 1974, *Vietnam studies: U.S. Army Engineers, 1965–1970.* US Department of the Army, Washington, 240 pp., pp. 95–104.
Porter, D.G., 1972, Bombing the dikes: Nixon's next option. *New Republic,* Washing-

ton, **166**(23), pp. 19–20.
Powers, E., 1890, *War and the Weather*, rev. ed. Edward Powers, Delavan, Wisconsin, 202 pp.
Prescott, J.A. & Pendleton, R.L., 1952, *Laterite and Lateritic Soils*. British Commonwealth Bureau of Soil Science, Harpenden, Technical Communication No. 47, 51 pp. + 7 ph.
Punte, C.L., Weimer, J.T., Ballard, T.A. & Wilding, J.L., 1962, Toxicological studies on *o*-chlorobenzylidene malononitrile. *Toxicology and Applied Pharmacology*, New York, **4**, pp. 656–62.

Quisenberry, K.S., 1970, *Herbicide Manual for Noncropland Weeds*, rev. ed. US Department of the Army, Washington, Technical Manual No. 5-629, [217] pp.

Radeleff, R.D., 1970, *Veterinary Toxicology*, 2nd ed. Lea and Febiger, Philadelphia, 352 pp.
Randal, J., 1967, Foe's sanctuary hit by fire bombs. *New York Times*, 19 January 1967, pp. 1, 3.
Rao, V.P., Ghani, M.A., Sankaran, T. & Mathur, K.C., 1971, *Review of the Biological Control of Insects and Other Pests in South-east Asia and the Pacific Region*. British Commonwealth Institute of Biological Control, Trinidad, West Indies, Technical Communication No. 6, 149 pp.
Rasmussen, D.I., 1941, Biotic communities of Kaibab plateau, Arizona. *Ecological Monographs*, US, **11**, pp. 229–75.
Red Cross, International Committee of the, 1973, *Weapons that may Cause Unnecessary Suffering or have Indiscriminate Effects*. International Committee of the Red Cross, Geneva, 72 pp.
Rees, D., 1964, *Korea: The Limited War*. St. Martin's, New York, 511 pp. + 15 pl.
Reinhart, K.G., 1973, *Timber-harvest Clearcutting and Nutrients in the Northeastern United States*. US Forest Service, Washington, Research Note No. NE-170, 5 pp.
Reinhold, R., 1972, U.S. attempted to ignite Vietnam forests in '66–67. *New York Times*, 21 July 1972, pp. 1–2; 22 July 1972, p. 5.
Rexrode, C.O. & Lockyer, J.W., 1974, *Laboratory Assay of Cacodylic Acid and ®Meta-Systox-R on* Scolytus multistriatus and Pseudopityophthorus sp. US Forest Service, Washington, Research Note No. NE-190, 4 pp.
Richards, P.W., 1952, *Tropical Rain Forest: An Ecological Study*. Cambridge University Press, London, 450 pp. + 15 pl. + 4 figs.
Richards, P.W., 1970, *Life of the Jungle*. McGraw-Hill, New York, 232 pp.
Richards, P.W., 1971, Some problems of nature conservation in the tropics. *Bull. Jardin Bot. Nat'l Belg.*, **41**, 173–87.
Richards, P.W., 1973, Tropical rain forest. *Scientific American*, New York, **229**(6), pp. 58–67, 148.
Richardson, J.M., 1960, Herbicidal control of rain forest regrowth. *Papua and New Guinea Agricultural Journal*, Port Moresby, **13**(2), pp. 66–69.
Ripley, S.D. *et al.*, 1964, *Land and Wildlife of Tropical Asia*. Time-Life Books, New York, 200 pp.
Rivers, L.M., (ed.), 1970, *Hearings on Military Posture . . .* US House of Representatives, Washington, Committee on Armed Services, Publication No. 91-53, 2 vols. (pp. 6811–7907+I–II and 7908–8667+I–XVII.)
Robinson, C.A., Jr., 1973, Special report: fuel air explosives: services ready joint development plan. *Aviation Week and Space Technology*, New York, **98**(8), pp. 42–46.
Rogers, B.W., 1974, *Vietnam Studies: Cedar Falls – Junction City: A Turning Point*. US Department of the Army, Washington, 172 pp.
Rogers, H.C.B., 1971, *Artillery through the Ages*. Seeley Service & Company, London, 230 pp. + 29 pl.

Rollet, B., 1962, *[Forest Inventory East of the Mekong.]* (In French) Food and Agriculture Organization of the United Nations, Rome, Report No. 1500, 184 pp. + [14] pl.

Romancier, R.M., 1965, *2,4-D, 2,4,5-T, and Related Chemicals for Woody Plant Control in the Southeastern United States.* Georgia Forest Research Council, Macon, Georgia, Report No. 16, 46 pp.

Rose, H. & Rose, S., 1972, CS gas: an imperialist technology. In: Browning, F. & Forman, D., (eds.), *Wasted Nations.* Harper and Row, New York, 346 pp., pp. 38–49.

Rosenblad, E., 1974, *Prohibited Weapons: Treaties and Bibliography.* Royal Swedish Library Bibliographical Institute, Stockholm, Documentation and Data Series No. 6, 50+2 pp.

Ross, P., 1974, *Effects of Herbicides in South Vietnam. B[14]. Effects of Herbicides on the Mangrove of South Vietnam.* US National Academy of Sciences, Washington, 33 pp.

Rowe, V.K. & Hymas, T.A., 1954, Summary of toxicological information on 2,4-D and 2,4,5-T type herbicides and an evaluation of the hazards to livestock associated with their use. *American Journal of Veterinary Research,* Chicago, **15**, pp. 622–29.

Rudd, R.L., 1964, *Pesticides and the Living Landscape.* University of Wisconsin Press, Madison, Wisconsin, 320 pp.

Sanders, H.O., 1970, Toxicities of some herbicides to six species of freshwater crustaceans. *Journal of the Water Pollution Control Federation,* Washington, **42**, pp. 1544–50.

Sargent, F., II. 1967, Dangerous game: taming the weather. *Scientist and Citizen,* Saint Louis, **9**, pp. 81–88, 96.

Sauer, C.O., 1950, Grassland climax, fire, and man. *Journal of Range Management,* Denver, **3**, pp. 16–21.

Savage, J.W., Jr., 1973, Tough row to plow. *Engineer,* Fort Belvoir, Virginia, **3**(4), pp. 12–14.

Schedler, P.W., 1968, Herbicides in tropical tree crops. *World Crops,* London, **20**(5), pp. 20–24.

Schindler, D. & Toman, J., 1973, *Laws of Armed Conflicts: A Collection of Conventions, Resolutions and other Documents.* A.W. Sijthoff, Leiden, 795 pp.

Schubert, O.E., 1967, Can activated charcoal protect crops from herbicide injury? *Crops and Soils,* Madison, Wisconsin, **19**(9), pp. 10–11.

Schultze, E., 1915, [French forests and the war.] (In German) *Zeitschrift für Forst- und Jagdwesen,* **47**, pp. 497–512.

Schuth, C.K., Isensee, A.R., Woolson, E.A. & Kearney, P.C., 1974, Distribution of ^{14}C and arsenic derived from [^{14}C]cacodylic acid in an aquatic ecosystem. *Journal of Agricultural and Food Chemistry,* Washington, **22**, pp. 999–1003.

Seigworth, K.J., 1943, Ducktown: a postwar challenge. *American Forests,* Washington, **49**, pp. 521–23, 558.

SIPRI. *See* Stockholm International Peace Research Institute.

Shapley, D., 1972a, Herbicides: DOD study of Viet use damns with faint praise. *Science,* Washington, **177**, pp. 776–79.

Shapley, D., 1972b, Rainmaking: rumored use over Laos alarms arms experts, scientists. *Science,* Washington, **176**, pp. 1216–20.

Shapley, D., 1972c, Technology in Vietnam: fire storm project fizzled out. *Science,* Washington, **177**, pp. 239–41.

Shapley, D., 1974, Weather warfare: Pentagon concedes 7-year Vietnam effort. *Science,* Washington, **184**, pp. 1059–61.

Sharp, U.S.G. & Westmoreland, W.C., (n.d.), *Report on the War in Vietnam (as of 30 June 1968).* US Government Printing Office, Washington, 347 pp. + maps.

Sheets, T.J. & Harris, C.I., 1965, Herbicide residues in soils and their phytotoxicities to crops grown in rotations. *Residue Reviews,* **11**, pp. 119–40.

Sivarajasingham, S., Alexander, L.T., Cady, J.G. & Cline, M.G., 1962, Laterite. *Advances in Agronomy,* **14**, pp. 1–60.
Sjödén, P.-O. & Söderberg, U., 1972, Sex-dependent effects of prenatal 2,4,5-trichlorophenoxy-acetic acid on rats open-field behavior. *Physiology and Behavior,* Fayetteville, New York, **9**, pp. 357–60.
Sjödén, P.-O. & Söderberg, U., 1975, Long-lasting effects of prenatal 2,4,5-trichlorophenoxyacetic acid on open-field behavior in rats: pre- and postnatal mediation. *Physiological Psychology,* Austin, Texas, **3**(2), pp. 175–78.
Smith, H.H. *et al.,* 1967a, *Area Handbook for North Vietnam.* US Department of the Army, Washington, Pamphlet No. 550–57, 494 pp.
Smith, H.H. *et al.,* 1967b, *Area Handbook for South Vietnam.* US Department of the Army, Washington, Pamphlet No. 550–55, 510 pp.
Smith, R.J., Jr., & Shaw, W.C., 1966, *Weeds and their Control in Rice Production.* US Department of Agriculture, Washington, Agriculture Handbook No. 292, 64 pp.
Sobolev, S.S., 1945, [Soil Erosion in the Territory of the Ukrainian SSR and the Struggle Against it after the Reconstruction of Agriculture Following the German Occupation.] (In Russian with English summary) *Pochvovedenie,* Moscow, **1945,** pp. 216–21 + 1 map.
Sobolev, S.S., 1947, Protecting the soils in the U.S.S.R. *Journal of Soil and Water Conservation,* U.S., **2**, pp. 123–32.
Soerjani, M., 1970, *Alang-alang:* Imperata cylindrica *(L.) Beauv. (1812): pattern of growth as related to its problem of control.* University of Gadjah Mada, Bogor, Indonesia, Ph.D. thesis, 88 pp. (Also: *Biotrop Bulletin,* Bogor, Indonesia, **1970**(1), 88 pp.)
Sopper, W.E. & Lull, H.W., (eds.), 1967, *Forest Hydrology.* Pergamon Press, New York, 813 pp.
Spencer, E.Y., 1968, *Guide to the Chemicals Used in Crop Protection,* 5th ed. Canada Department of Agriculture, Ottawa, Publication No. 1093, 483 pp.
Spurr, S.H. & Barnes, B.V., 1973, *Forest Ecology,* 2nd ed. Ronald Press, New York, 571 pp.
Stahler, L.M. & Whitehead, E.I., 1950, Effect of 2,4-D on potassium nitrate levels in leaves of sugar beets. *Science,* Washington, **112**, pp. 749–51.
Stark, N., 1971, Nutrient cycling. I. Nutrient distribution in some Amazonian soils. II. Nutrient distribution in Amazonian vegetation. *Tropical Ecology,* Varanasi, India, **12**, pp. 24–50, 177–201.
State, US Department of, 1963a, *Democratic Republic of Viet-Nam (North Viet-Nam): A Bibliography.* US Department of State, Washington, External Research Paper No. 142, 21 pp.
State, US Department of, 1963b, *Republic of Vietnam (South Vietnam): A Bibliography.* US Department of State, Washington, External Research Paper No. 143, 25+22 pp.
Stecher, P.G. *et al.,* (eds.), 1968, *Merck Index: An Encyclopedia of Chemicals and Drugs,* 8th ed. Merck, Rahway, New Jersey, 1713 pp.
Steele, B., 1968, Chemical weed control in tropical tree crops. *Pans,* London, **14**(3), pp. 250–54.
Stelzer, M.J., 1970, Mortality of *Ips lecontei* attracted to ponderosa pine trees killed with cacodylic acid. *Journal of Economic Entomology,* US, **63**, pp. 956–59.
Sterba, J.P., 1970, G.I.'s in Vietnam stir duststorms. *New York Times,* 19 April 1970, p. 24.
Sterle, J.R., 1967, *LeTourneau Tree Crusher.* American Pulpwood Association, New York, Technical Release No. 67-R-54, 5 pp.
Stevens, P.H., 1965, *Artillery Through the Ages,* F. Watts, New York, 197 pp. + ph.
Stidd, C.K., 1968, *Local Moisture and Precipitation,* rev. ed. University of Nevada Desert Research Institute, Reno, Preprint Series No. 45A, 34 pp. + 5 figs.

Stillman, E., 1972, Smart bombs and dumb strategy. *Saturday Review,* New York, **55**(31), pp. 27–32.
Stockholm International Peace Research Institute (SIPRI), 1969, *SIPRI Yearbook of World Armaments and Disarmament 1968/69.* Almqvist & Wiksell, Stockholm, 440 pp.
Stockholm International Peace Research Institute (SIPRI), 1971–1975, *The Problem of Chemical and Biological Warfare.* Almqvist & Wiksell, Stockholm, 6 vols. (395+420+194+412+287+308 pp.)
Stockholm International Peace Research Institute (SIPRI), 1975, *Incendiary Weapons.* Almqvist & Wiksell, Stockholm, 255 pp.
Stone, C.D., 1972, Should trees have standing?: toward legal rights for natural objects. *Southern California Law Review,* Los Angeles, **45**, pp. 450–501. [Also: William Kaufmann, Los Altos, California, 103 pp., 1974.]
Studer, T.A., 1968–1969, Weather modification in support of military operations. *Air University Review,* Maxwell Air Force Base, Alabama, **20**(6), pp. 44–50.
Stuttard, J.C. *et al.,* 1943, *Indo-China.* British Admiralty, Naval Intelligence Division, Geographic Handbook Series No. BR510, 535 pp. + 110 pl. + 1 map.
Sus, N.I., 1944, [Agricultural and forest amelioration and science in the war and post-war periods.] (In Russian) *Bulletin of the Institute of Grain Economy of Southeastern USSR,* Saratov, **1944**(3), pp. 3–14.
Swanson, C.R. & Shaw, W.C., 1954, Effect of 2,4-dichlorophenoxyacetic acid on the hydrocyanic acid and nitrate content of sudan grass. *Agronomy Journal,* Madison, Wisconsin, **46**, pp. 418–21.
Swanson, C.W., 1975, Reforestation in the Republic of Vietnam. *Journal of Forestry,* Washington, **73**, pp. 367–71.
Swiggart, R.C., Whitehead, C.J., Jr., Curley, A. & Kellogg, F.E., 1972, Wildlife kill resulting from the misuse of arsenic acid herbicide. *Bulletin of Environmental Contamination and Toxicology,* West Berlin, **8**(2), pp. 122–28.
Sylva, D.P. de & Michel, H.B., 1974, *Effects of Herbicides in South Vietnam. B[15]. Effects of Mangrove Defoliation on the Estuarine Ecology and Fisheries of South Vietnam.* US National Academy of Sciences, Washington, 126 pp.
Taborsky, O. & Thuronyi, G., 1960, Annotated bibliography on weather modification. *Meteorological and Geoastrophysical Abstracts,* Boston, **11**, pp. 2181–415.
Taborsky, O. & Thuronyi, G., 1962, Annotated bibliography on weather modification and microphysics of clouds. *Meteorological and Geoastrophysical Abstracts,* Boston, **13**, pp. 702–62.
Takman, J., (ed.), 1967, *[Napalm.]* (In Swedish) Rabén & Sjögren, Stockholm, 189 pp.
Talbot, L.M. & Talbot, M.H., (eds.), 1968, *Conservation in Tropical South East Asia.* International Union for Conservation of Nature and Natural Resources, Morges, Switzerland, Publication New Series No. 10, 550 pp. + 2 pl.
Tân, Phan Huy., 1971, *[Defoliation with Chemicals.]* (In Vietnamese) Republic of Vietnam Directorate of Waters and Forests, Saigon, 6 pp.
Taylor, B.W., 1957, Plant succession on recent volcanoes in Papua. *Journal of Ecology,* Oxford, **45**, pp. 233–43 + 1 pl.
Tempany, H.A., 1951, *Imperata* grass: a major menace in the wet tropics. *World Crops,* London, **3**, pp. 143–46.
Thom, C. & Raper, K.B., 1932, Arsenic fungi of Gosio. *Science,* Washington, **76**, pp. 548–50.
Thomas, R.E., Cohen, J.M. & Bendixen, T.W., 1964, *Pesticides in Soil and Water: An Annotated Bibliography.* US Public Health Service, Washington, Publication No. 999-WP-17, 90 pp.
Thomas, W.L., Jr. *et al.,* (eds.), 1956, *Man's Role in Changing the Face of the Earth.* University of Chicago Press, Chicago, 1193 pp.

Thuronyi, G., 1963, Annotated bibliography on weather modification and microphysics of clouds (supplement). *Meteorological and Geoastrophysical Abstracts,* Boston, **14**, pp. 144–244.

Thuronyi, G., 1964, Recent literature on weather and climate modification. *Meteorological and Geoastrophysical Abstracts,* Boston, **15**, pp. 1518–53.

Tietjen, H.P., Halvorson, C.H., Hegdal, P.L. & Johnson, A.M., 1967, 2,4-D herbicide, vegetation, and pocket gopher relationships Black Mesa, Colorado. *Ecology,* U.S., **48**, pp. 634–43.

Time, 1968, Shrinking sanctuary. *Time,* New York, **91**(17), p. 28.

Times, New York, 1970, . . . and a plea to ban 'ecocide'. *New York Times,* 26 February 1970, p. 38.

Toumey, J.W. & Korstian, C.F., 1947, *Foundations of Silviculture upon an Ecological Basis,* 2nd ed. John Wiley, New York, 468 pp.

Tregonning, K.G., 1969, *Southeast Asia: A Critical Bibliography.* University of Arizona Press, Tucson, Arizona, 103 pp.

Trichell, D.W., Morton, H.L. & Merkle, M.G., 1968, Loss of herbicides in runoff water. *Weed Science,* Urbana, Illinois, **16**, pp. 447–49.

Troendle, C.A., 1970, *Comparison of Soil-moisture Loss from Forested and Clearcut Areas in West Virginia.* US Forest Service, Washington, Research Note No. NE-120, 8 pp.

Trousdell, K.B. & Hoover, M.D., 1955, Change in ground-water level after clearcutting of loblolly pine in the Coastal Plain. *Journal of Forestry,* Washington, **53**, pp. 493–98.

Truman, R., 1961–1962, Eradication of mangroves. *Australian Journal of Science,* Sydney, **24**, pp. 198–99.

Tschirley, F.H., 1968, *Research Report: Response of Tropical and Subtropical Woody Plants to Chemical Treatments.* US Agricultural Research Service, Beltsville, Maryland, Publication No. CR-13-67, 197 pp.

Tschirley, F.H., 1969, Defoliation in Vietnam. *Science,* Washington, **163**, pp. 779–86.

Tucker, R.K. & Crabtree, D.G., 1970, *Handbook of Toxicity of Pesticides to Wildlife.* US Fish and Wildlife Service, Washington, Resource Publication No. 84, 131 pp.

Tukey, J.W. *et al.*, 1965, *Restoring the Quality of Our Environment.* The White House, Washington, 317 pp.

Tung, Thái Công, 1967, *Natural Environment and Land Use in South Vietnam.,* 2nd ed. Republic of Vietnam Ministry of Agriculture, Saigon, 156 pp. + 3 maps.

Union of Soviet Socialist Republics, 1974, *Prohibition of Action to Influence the Environment and Climate for Military and other Purposes Incompatible with the Maintenance of International Security, Human Well-being and Health.* United Nations, New York, Document No. A/C.1/L.675, 2+5 pp.

Union of Soviet Socialist Republics, 1975, *Draft Convention on the Prohibition of Military or any other Hostile Use of Environmental Modification Techniques.* Conference of the Committee on Disarmament, Geneva, Document No. CCD/471, 3 pp.

United Nations, 1945, Charter, In: Russell, R.B., 1958, *History of the United Nations Charter.* Brookings Institution, Washington, 1140 pp., pp. 1035–53.

United Nations War Crimes Commission, 1948, *History of the United Nations War Crimes Commission and the Development of the Laws of War.* His Majesty's Stationery Office, London, 592 pp.

United States of America, 1975, *Draft Convention on the Prohibition of Military or any other Hostile Use of Environmental Modification Techniques.* Conference of the Committee on Disarmament, Geneva, Document No. CCD/472, 3 pp.

Valder, S.M., 1972, *Insect Density and Diversity Studies on Test Area C-52A, Eglin AFB Reservation, Florida.* US Air Force Armament Laboratory, Eglin Air Force Base, Florida, Technical Note No. 72-4, 25 pp.

Valentine, J.P. & Bingham, S.W., 1974, Influence of several algae on 2,4-D residues in water. *Weed Science,* Urbana, Illinois, **22**, pp. 358–63.
VandenBorn, W.H., 1969, Picloram residues and crop production. *Canadian Journal of Plant Science,* Ottawa, **49**, pp. 628–29.
VanderEls, T., 1971, Irresistible weapon. *Military Review,* Fort Leavenworth, Kansas, **51**(8), pp. 80–90.
Vietnam Newsletter, 1971, Worst flood ever. *Vietnam Bulletin,* Washington, **6**(9), pp. 7–9.
Voigt, G.K., 1960, Alteration of the composition of rainwater by trees. *American Midland Naturalist,* **63**, pp. 321–26.
Waggoner, P.E., 1966, Weather modification and the living environment. In: Darling, F.F. & Milton, J.P., (eds.), *Future Environments of North America.* Natural History Press, Garden City, New York, 770 pp., pp. 87–98.
Wagner, R.H., 1974, *Environment and Man,* 2nd ed. W.W. Norton, New York, 528 pp.
Wagner, S.L. & Weswig, P., 1974, Arsenic in blood and urine of forest workers: as indices of exposure to cacodylic acid. *Archives of Environmental Health,* Chicago, **28**, pp. 77–79.
Walker, E.P. *et al.,* 1964, *Mammals of the World.* Johns Hopkins Press, Baltimore, 2 vols. (1500 pp.)
Walker, P.H. & Costin, A.B., 1971, Atmospheric dust accession in south-eastern Australia. *Australian Journal of Soil Research,* Melbourne, **9**, pp. 1–5.
Walsh, G.E., 1967, Ecological study of a Hawaiian mangrove swamp. In: Lauff, G.H., (ed.), *Estuaries.* American Association for the Advancement of Science, Washington, Publication No. 83, 757 pp., pp. 420–31.
Walsh, G.E., Barrett, R., Cook, G.H. & Hollister, T.E., 1973, Effects of herbicides on seedlings of the red mangrove, *Rhizophora mangle* L. *BioScience,* Washington, **23**, pp. 361–64.
War, US Department of, 1943, *Unexploded Bombs: Organization and Operation for Disposal.* US Department of War, Washington, Field Manual No. 9–40, 148 pp.
Ward, F.P., 1973, *Progress in Ecological Research at Edgewood Arsenal, Maryland: Fiscal Years 1971 and 1972.* US Army Edgewood Arsenal, Maryland, Special Publication No. 1100–13, 20 pp.
Warren, J.R., Graham, F. & Gale, G., 1951, Dominance of an actinomycete in a soil microflora after 2,4-D treatment of plants. *Phytopathology,* Saint Paul, Minnesota, **41**, pp. 1037–39.
Way, J.M., 1969, Toxicity and hazards to man, domestic animals, and wildlife from some commonly used auxin herbicides. *Residue Reviews,* **26**, pp. 37–62.
Weast, R.C., (ed.), 1974, *Handbook of Chemistry and Physics,* 55th ed. CRC Press, Cleveland, [2279] pp.
Weatherspoon, C.P. & Krusinger, A.E., 1974a, *Effects of Herbicides in South Vietnam. B[17]. Air-photo Inventory of the Rung-Sat.* US National Academy of Sciences, Washington, 11 pp.
Weatherspoon, C.P. & Krusinger, A.E., 1974b, *Effects of Herbicides in South Vietnam. B[18]. Air-photo Studies of the Rung-Sat.* US National Academy of Sciences, Washington, 25 pp.
Weimer, J.T., Ballard, T.A., Owens, E.J. & McNamara, B.P., 1970, *Toxicological Studies on the Herbicide "White" in Animals.* US Army Edgewood Arsenal, Maryland, Technical Report No. 4439, 35 pp.
Weisberg, B., (ed.), 1970, *Ecocide in Indochina: The Ecology of War.* Canfield Press, San Francisco, 241 pp. + 11 ph.
Wells, P.V., 1965, Scarp woodlands, Transported grassland soils, and concept of grassland climate in the Great Plains region. *Science,* Washington, **148**, pp. 246–49.
Went, F.W. & Stark, N., 1968, Mycorrhiza. *BioScience,* Washington, **18**, pp. 1035–39.

Westing, A.H., 1971a, Ecocide in Indochina. *Natural History,* New York, **80**(3), pp. 56–61, 88.
Westing, A.H., 1971b, Ecological effects of military defoliation on the forests of South Vietnam. *BioScience,* Washington, **21**, pp. 893–98.
Westing, A.H., 1971c, Forestry and the war in South Vietnam. *Journal of Forestry,* Washington, **69**, pp. 777–83.
Westing, A.H., 1971d, Leveling the jungle. *Environment,* Saint Louis, **13**(9), pp. 8–12.
Westing, A.H., 1971–1972a, Herbicides as agents of chemical warfare: their impact in relation to the Geneva Protocol of 1925. *Environmental Affairs,* Boston, **1**, pp. 578–86.
Westing, A.H., 1971–1972b, Herbicides in war: current status and future doubt. *Biological Conservation,* Barking, England, **4**, pp. 322–27.
Westing, A.H., 1972a, Herbicidal damage to Cambodia. In: Neilands, J.B. *et al., Harvest of Death: Chemical Warfare in Vietnam and Cambodia.* Free Press, New York, 304 pp., pp. 177–205.
Westing, A.H., 1972b, [Letter to the editor.] *Journal of Forestry,* Washington, **70**, p. 129.
Westing, A.H., 1972c, Super bomb. *American Report,* New York, **2**(45), p. 3.
Westing, A.H., 1972d, U.S. Food destruction program in South Vietnam. In: Browning, F. & Forman, D., (eds.), *Wasted Nations.* Harper and Row, New York, 346 pp., pp. 21–25.
Westing, A.H., 1973a, AAAS Herbicide Assessment Commission. *Science,* Washington, **179**, pp. 1278–79.
Westing, A.H., 1973b, Postwar visit to Hanoi. *Boston Globe,* **204**(85), p. A6.
Westing, A.H., 1974a, *Herbicides as Weapons: A Bibliography.* California State University, Los Angeles, Center for the Study of Armament and Disarmament, Political Issues Series, **3**(1), 36 pp.
Westing, A.H., 1974b, Postwar forestry in North Vietnam. *Journal of Forestry,* Washington, **72**, pp. 153–56.
Westing, A.H., 1974c, Proscription of ecocide: arms control and the environment. *Bulletin of the Atomic Scientists,* Chicago, **30**(1), pp. 24–27.
Westing, A.H. & Pfeiffer, E.W., 1972, Cratering of Indochina. *Scientific American,* New York, **226**(5), pp. 20–29, 138; (6), p. 7.
Wharton, C.H., 1966, Man, fire and wild cattle in north Cambodia. *Proceedings of the Tall Timbers Fire Ecology Conference,* US, **5**, pp. 23–65.
Wharton, C.H., 1968, Man, fire and wild cattle in southeast Asia. *Proceedings of the Tall Timbers Fire Ecology Conference,* US, **7**, pp. 107–67.
Whitaker, D.P. *et al.,* 1972, *Area Handbook for Laos,* 2nd ed. US Department of the Army, Washington, Pamphlet No. 550–58, 335 pp.
Whitaker, D.P. *et al.,* 1973, *Area Handbook for the Khmer Republic (Cambodia).* US Department of the Army, Washington, Pamphlet No. 550–50, 387 pp.
White, W.D., 1974, *U.S. Tactical Air Power: Missions, Forces, and Costs.* Brookings Institution, Washington, 121 pp.
Whitehead, E.I., Kersten, J. & Jacobsen, D., 1956, Effect of 2,4-D spray on the nitrate content of sugar beet and mustard plants. *Proceedings of the South Dakota Academy of Sciences,* **35**, pp. 106–10.
Whitehead, H.C. & Feth, J.H., 1964, Chemical composition of rain, dry fallout, and bulk precipitation at Menlo Park, California, 1957–1959. *Journal of Geophysical Research,* Washington, **69**, pp. 3319–33.
Whitney, C.R., 1969, Naval gunfire in Vietnam. *Ordnance,* Washington, **53**, pp. 602–606.
Whitney, C.R., 1971, Air war: the idea is not to kill civilians – if possible. *New York Times,* 18 April 1971, p. E2.
Whitney, C.R., 1972, B-52 relied upon more than troops to blunt foe's offensive in Vietnam. *New York Times,* 19 May 1972, p. 9.

Whittaker, R.H., 1967, Ecological implications of weather modification. In: Shaw, R.H., (ed.), *Ground Level Climatology*. American Association for the Advancement of Science, Washington, Publication No. 86, 395 pp., pp. 367–84.

Wilde, S.A., Steinbrenner, E.C., Pierce, R.S., Dosen, R.C. & Pronin, D.T., 1953, Influence of forest cover on the state of the ground water table. *Soil Science Society of America Proceedings*, Madison, Wisconsin, **17**, pp. 65–67.

Willard, C.J., 1950, Indirect effects of herbicides. *Proceedings of the North Central Weed Control Conference*, US, **7**, pp. 110–12.

Williams, K.T. & Whetstone, R.R., 1940, *Arsenic Distribution in Soils and its Presence in Certain Plants*. US Department of Agriculture, Washington, Technical Bulletin No. 732, 20 pp.

Williams, L., 1965, *Vegetation of Southeast Asia: Studies of Forest Types, 1963–1965*. US Agricultural Research Service, Beltsville, Maryland, Publication No. CR 49-65, 302 pp.

Williams, L., 1967, *Forests of Southeast Asia, Puerto Rico, and Texas*. US Agricultural Research Service, Beltsville, Maryland, Publication No. CR 12-67, 410 pp.

Wing, L.W., 1951, *Practice of Wildlife Conservation*. John Wiley, New York, 412 pp.

Winters, R.K., 1974, *Forest and Man*. Vantage Press, New York, 393 pp.

Wood, H.B., 1971, Land use effects on the hydrologic characteristics of some Hawaii soils. *Journal of Soil and Water Conservation*, US, **26**, pp. 158–60.

Wood, J.M., 1974, Biological cycles for toxic elements in the environment. *Science*, Washington, **183**, pp. 1049–52.

Woodwell, G.M., 1970, Effects of pollution on the structure and physiology of ecosystems. *Science*, Washington, **168**, pp. 429–33.

Woolson, E.A. & Kearney, P.C., 1973, Persistence and reactions of ^{14}C-cacodylic acid in soils. *Environmental Science and Technology*, Washington, **7**, pp. 47–50.

Woudt, B.D. van't & Uehara, G., 1961, *Erosion Behavior and Control of a Stripmined Latosolic Soil*. Hawaii Agricultural Experiment Station, Honolulu, Technical Bulletin No. 46, 36 pp.

Wright, Q., 1965, *Study of War: With a Commentary on War Since 1942*, 2nd ed. University of Chicago Press, 1637 pp.

Wrigley, G., 1961, Advances in the use of agricultural chemicals in tropical agriculture. *Tropical Agriculture*, Trinidad, **38**, pp. 271–82.

Wulff, T., Janzon, B., Ohlson, L.-O., Petré, T. & Rybeck, B., 1973, *Conventional Weapons, their Deployment and Effects from a Humanitarian Aspect: Recommendations for the Modernization of International Law*. Swedish Ministry for Foreign Affairs, Stockholm, 182 pp.

Yamamoto, T. & Duffy, P., 1963, *Water Storage Capacities of Soil under Four Different Land Uses in Hawaii*. US Forest Service, Washington, Research Note No. PSW-5, 3 pp.

Yoda, K. & Kira, T., 1969, Comparative ecological studies on three main types of forest vegetation in Thailand. V. Accumulation and turnover of soil organic matter, with notes on the altitudinal soil sequence on Khao (Mt.) Luang, peninsular Thailand. *Nature and Life in Southeast Asia*, Tokyo, **6**, pp. 83–110 + 2 pl.

Young, A.L., 1974, *Ecological Studies on a Herbicide-equipment Test Area (TA C-52A) Eglin AFB Reservation, Florida*. US Air Force Armament Laboratory, Eglin Air Force Base, Florida, Technical Report No. 74–12, 141 pp.

Young, A.L. & Wolverton, B.C., 1970, *Military Herbicides and Insectides*. US Air Force Armament Laboratory, Eglin Air Force Base, Florida, Technical Note No. 70-1, 59 pp.

Youngberg, C.T., 1965, Silvicultural benefits from brush. *Society of American Foresters Proceedings*, Washington, **1965**, pp. 55–59.

Younge, O.R. & Moomaw, J.C., 1960, Revegetation of stripmined bauxite lands in Hawaii. *Economic Botany*, New York, **14**, pp. 316–30.

Zabel, R.A. & O'Neil, F.W., 1957, Toxicity of arsenical compounds to microorganisms. *Tappi*, New York, **40**, pp. 911–14.

Zablocki, C.J., (ed.), 1970, *Chemical-Biological Warfare: U.S. Policies and International Effects. [I.] Hearings. [II.] Report.* US House of Representatives, Washington, Committee on Foreign Affairs, 513+41 pp.

Zimin, V.B., 1971, [Influence of arboricides on useful fauna.] (In Russian) In: Shubin, V.I. & Morozova, R.M., (eds.), *[Fertilizers and Herbicides in Forestry in the North of European Russia.]* Izdatel'stvo Nauka, Leningrad, 175 pp., pp. 92–97.

Zinke, P.J., 1974, *Effects of Herbicides in South Vietnam. B[19]. Effect of herbicides on Soils of South Vietnam.* US National Academy of Sciences, Washington, 39 pp.